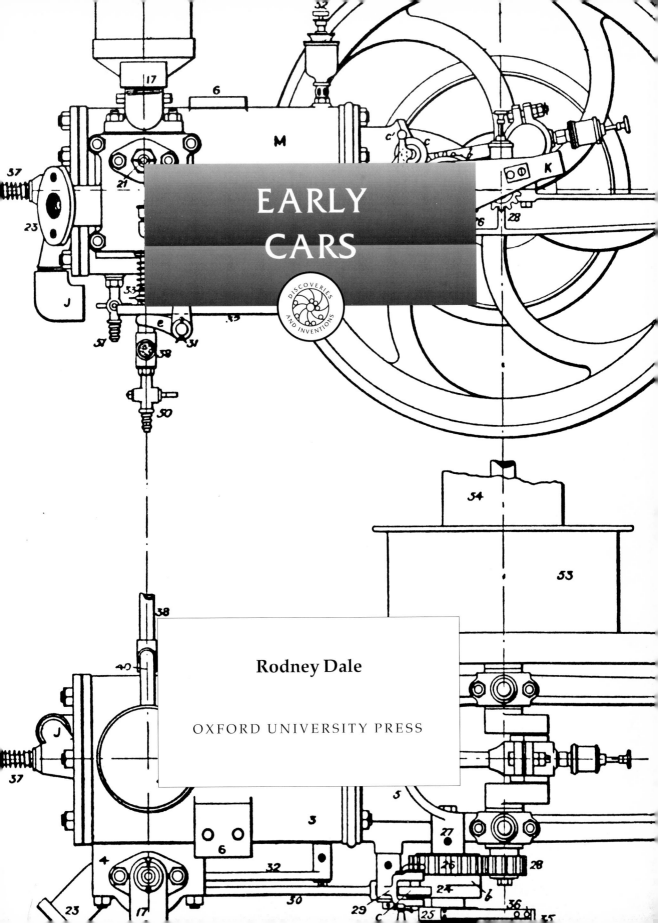

EARLY CARS

DISCOVERIES AND INVENTIONS

Rodney Dale

OXFORD UNIVERSITY PRESS

Photographic Acknowledgements

All illustrations in this book have been taken from out-of-copyright
material in the British Library's collections, with the exception of those
which carry a credit line in the accompanying caption. Patents are
numbered and dated for ease of reference.

OXFORD UNIVERSITY PRESS

Oxford   New York   Toronto
Delhi   Bombay   Calcutta   Madras   Karachi
Kuala Lumpur   Singapore   Hong Kong   Tokyo
Nairobi   Dar es Salaam   Cape Town
Melbourne   Auckland   Madrid
and associated companies in
Berlin   Ibadan

Library of Congress Cataloging-in-Publication Data

Dale, Rodney, 1933-
    Early cars / Rodney Dale.
    64p. 24.6 x 18.9 cm. — (Discoveries and Inventions)
    Includes bibliographical references and index.
    Summary: Examines the invention, development, and technology of
the earliest automobiles, discussing their vital components and how
they were refined over the years.
    ISBN 0-19-521002-6. — ISBN 0-19-521006-9 (pbk.)
    1. Antique and classic cars — History — Juvenile literature.
2. Automobile travel — History — Juvenile literature.
[1. Automobiles — History.] I. Title. II. Series.
TL 15.D27 1993
629.222 — dc20   93-3660
                                                  CIP
                                                  AC

Designed by Roger Davies
Set in Palatino on Ventura
Printed in Italy

Contents

Early self-moving carriages

What is a car?

A 'car' is a type of carriage – in the 14th century a name for a two-wheeled wagon for transporting heavy loads. Later, the word 'car' took on a more poetically majestic meaning as the vehicle by which the heavenly bodies – particularly the sun – rode across the sky.

What we now think of as a motor car, or automobile, was originally called a horseless carriage, for obvious reasons. The term motor car was introduced in 1895 by the pioneering F R Simms, who saw that perpetuating the idea of the horseless carriage was not very forward looking; a Motor Car Club was formed in England the following year.

In this book the word 'car' is a wheeled vehicle propelled on the roads by some type of internal power, provided first by animals (including man), then by steam, and currently by internal combustion.

Muscle power

There were many attempts, actual and reputed, to propel vehicles by muscle power. There are ancient Greek accounts of a 'triumphal wagon of Athens moved along by men carried therein'. The 13th-century philosopher Roger Bacon suggested the possibility of making carriages move 'without animals', but the idea

The author's grandmother and great uncle on the Belgian Minerva in 1908.

The Nuremberg carriage of 1649. Since it could carry several passengers, the two men inside must have found propelling it hard work – especially when they had to pump water and blow the trumpets while going uphill, all at the same time.

needs no real feat of imagination; it is as if we were to suggest cars moving 'without petrol'. Leonardo da Vinci (1452–1519) sketched a self-moving car – but he sketched many things which were never built.

European travellers in China in the 15th and 16th centuries reported seeing carriages propelled by rowing and by sails, but it was not until the 17th century that attempts to build automotive carriages were made in Europe.

In 1618 David Ramsey, Page of the Bedchamber to James I, and Thomas Wildgosse, gentleman, applied for a patent for a machine which would, among other things, plough without horses or oxen; they spoke also of boats which went without sails. In 1625, John Marshall patented 'a cart of 15cwt [750 kg] to carry a great burden without help of horses', presenting his petition to King Charles I at an audience in Whitehall.

Sir Isaac Newton (1642–1727) is supposed to have made a manumotive (literally, 'moved by hand') carriage while at school at Grantham in about 1655, but this may be one of the many wondrous inventions customarily attributed with hindsight to a great man. Of it, Sir David Brewster wrote: 'The mechanical carriage which Sir Isaac is said to have invented was a four-wheeled vehicle and was moved with a handle or winch wrought by the person who sat in it. We can find no distinct information respecting its construction or use, but it must have resembled a Merlin's chair which is fitted only to move on the smooth surface of a floor,

and not to overcome the inequalities of a common road.' The 'Merlin's chair' was a wheeled invalid chair named after a French mathematical instrument maker who came to London in 1760.

Several other attempts were made to improve road vehicles: Mr Potter's cart with legs instead of wheels (1633), and a new carriage, invented by the Duc de Roanez, presented to the Royal Society as a 'scheme of an instrument to walk upon the land or water with swiftness after the manner of the wheel of a crane.'

In 1665, the diarist John Evelyn wrote: 'On my return I called at Durdan's where I found Dr Wilkins, Sir William Petty and Mr Hooke contriving chariots, new rigging for ships, the wheele for one to run races in, and other mechanical inventions.' However, this sounds more like three philosophers having wide-ranging flights of fancy than a serious design session.

The Nuremberg carriage

A celebrated mechanician called Johan Hautsch of Nuremberg in Germany built an ornate carriage in 1649. It is thought to have been worked by two men concealed inside, who turned the rear axle by means of handles. It is reported to have gone up and down hills, and steered around corners, and stopped and started as desired, attaining a speed of 2,000 paces an hour (which isn't very fast). It could carry several passengers, and a dragon in front could spout out a stream of water to clear a way through a crowd. This might have been superfluous, because the dragon could turn its eyes to and fro with great rapidity. If this (and the water) didn't frighten people out of the way, the angels, mounted one on each side of the carriage, sounded their trumpets. Hautsch sold his first carriage to the Crown Prince of Sweden and later built another for the King of Denmark.

The Scandinavian crowds assembling to cheer the carriages as they crawled by must have been at least as hardy and patient as the men inside struggling to propel the great mass, heavy with water for the dragon, as they applied their lips to the angels' trumpets.

Richard's and other carriages

Towards the end of the 17th century, Elie Richard, a physician and scientist of La Rochelle, France, built a successful carriage propelled by pedals. The motive power was a servant in the box behind, who caused the wheels to rotate by means of treadles as shown in the diagram, while his master had the infinitely preferable task of steering, and admiring the passing show. It was later suggested that, if the treadles were brought to the front, the passenger himself might propel the carriage: 'a machine of this kind will afford a salutary recreation in a garden or park, or on any plain ground, but on a rough or deep road must be attended with more pain than pleasure.'

It is no surprise to find carriages propelled by humans being built from time to time. Edmund Cartwright (1743–1823), the inventor of the power loom, believed that 'an able bodied man can exert the power of a horse,' and thought that, if he were to live a few

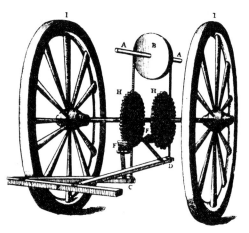

Richard's carriage, treadled by a servant in the box behind. The downstroke of the treadles provided forward motion; freewheels allowed them to rise again.

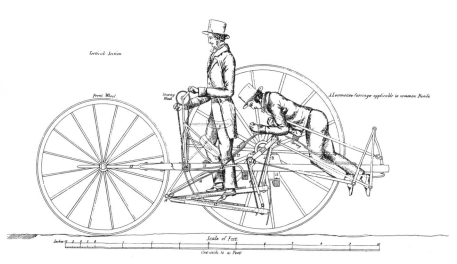

Like Richard's carriage, Bramley & Parker's 'locomotive carriage applicable to common roads' (British patent no 6027 of 1830) also used treadles and freewheels. There was one model for the solo locomotist, and another for up to three (two shown).

Snowden's carriage was propelled by horses walking round and round inside. It would need a road of some width to accommodate it.

years longer, carriages of every description would be travelling the public roads without the aid of horses. A few years later, in a letter dated 1822, Cartwright wrote: 'I have completed my invention of a carriage to go without horse, which I call a Centaur carriage. Two men took a cart from my house (cart and its load weighing 16 cwt [812kg]) a distance of 27 miles [43.5km] in a day, and up two very long and steep hills. Since then I have greatly improved upon it. It is now so constructed that I can give it what speed I please.' Cartwright's carriage had a pair of treadles and cranks worked alternately by the feet of the driver.

There was an interest in such carriages, apparently in the North of England, at the beginning of the 19th century. Little is known of Mr Bain's carriage of 1819, except that it was to be worked by pushing legs operated by treadles. A carriage propelled by treadles was

built at Sunderland in 1827, and is said to have carried seven people. Three years later, Isaac Brown of Bingley in Yorkshire made a 'wonderful wooden horse' which drew a gig over a mile in six minutes (16 kph). And soon after that, at about the time of the great steam-carriage boom (see p 13), a scheme was suggested for establishing a system of public conveyances in London to be worked by humans – perhaps even by the passengers themselves!

George Stephenson, 'father of the railways', himself proposed such a scheme. He wrote to the locomotive builder Timothy Hackworth, about Brandreth's horse-powered *Cyclopede*, claiming its invention, and that he had contemplated using it on the Canterbury & Whitstable railway. *Cyclopede* had been entered for the 'Rainhill Trials' – the Liverpool & Manchester Railway test in 1829, to select locomotives, but – apart from being disqualified – it was incapable of speeds higher than five or six miles per hour (9.7 kph) – even on rails.

W F Snowden's carriage of 1824 has an upper storey for the passengers, and a lower storey for goods and for the driving force – the horses, which walk round and round on a circular track. The horses are yoked to radial arms on a central vertical shaft which turns the axles of the road wheels via toothed gearing. It could have been suitable only for the broadest of thoroughfares.

Humans and horses tire; in the absence of steam, some other mechanical source of power was sought. Types of spring motors were suggested, but the problem is the relatively enormous weight of metal to be carried in order to store a sufficient amount of energy. In 1760, M Genevois – a French inventor of a sailing carriage – thought of installing a series of springs to drive it if the wind should fail.

The idea of clockwork vehicles persisted. Various ways of winding the springs of the motors were suggested – for example a steam engine or a gunpowder engine – even after the gas engine was well established. In 1870 a clockwork omnibus was tried on the streets of New Orleans. In 1873, D H Leveaux of Brook Green, London proposed using a series of springs under the floor of a car, to be wound up by stationary steam engines placed at suitable intervals along the road. A company was set up to work the system, and constructed a number of motors of up to two horse power. What happened then is not stated.

The power of the wind

Another source of power – excellent as long as it lasts, but unpredictable – is the wind. Simon Stevin (1548–1620) of Bruges, Belgium, mathematician and mechanician, built two carriages driven by wind. Another 'windmill carriage' was built by a Mr Wilkins in 1648. Towards the end of the 17th century, Sir Humphrey Mackworth who, according to a contemporary account in an 'Essay on Mines' was renowned for his

sailing-wagons for the cheap carriage of his coal to the water side whereby one horse does the work of ten at all times but when any wind is stirring (which is seldom wanting near the sea) one man and a small sail does the work of 20 ... And I believe, he is the first gentleman, in this part of the world, that hath set up sailing-engines on land, driven by the wind, not for any curiosity, or vain applause, but for real profit, ...

Stevin's sailing carriage of 1600 relied on the wind. It was probably more of a novelty than a practicable means of transport.

Mackworth's sailing wagons are almost outside our scope, for they ran on rails (which greatly reduces the friction), but we include them here because, far from being a fantasy, they worked, and worked well.

In 1834, a ship's captain at Llanelli in South Wales made a sailing carriage which was reported to have covered a distance of three miles (4.8 km) – including a steep incline – in seven and a half minutes (38.4 kph). Of many attempts in America the *Scientific American* (1878) published an illustrated description of a sailing car devised by Mr C J Bascombe of the Kansas Pacific Railroad. It had been in use for three years for conveying repair and maintenance gangs to pumps, telegraph lines and so on along the line of the railway at an average speed of 20 mph (32 kph). The drawing must have been made to show the effects of a strong breeze, because the occupants of the car appear to be exerting their full strength in holding on.

Viney and Pocock's charvolant or kite carriage of 1827 was furnished with a rear platform to carry a pony which could provide power if the wind should fail. A kite carriage built in about 1826 made the journey from Bristol to London, and was frequently to be seen in Hyde Park and in the northern suburbs of London. It is recorded that the carriage attained a speed of 20 mph (32 kph) on several occasions, even on heavy roads. On one occasion the mile was covered in 2.75

Viney & Pocock's kite carriages, or charvolants, travelling in various directions with the same wind.

minutes (35 kph). In answer to questions about the power of his kites, Mr Pocock stated that a man of moderate strength could just hold a kite 12 feet (3.65m) high, with an area of 49 square feet (4.5m^2) against a wind blowing at 20 mph (32 kph) and that, with a rather stronger wind, the kite broke a string capable of sustaining a weight of 200 pounds (90.7 kg).

The power of steam

There were many early attempts to apply steam power to road vehicles. The 'Æliopile' is a primitive form of steam turbine and is said to have been used for driving a carriage built by Father Verbiest (born near Courtrai 1623; died in China 1688) when he was in China. It is hard to believe that enough power could be derived from this device; the more it was geared down, the greater would be the frictional losses in the system.

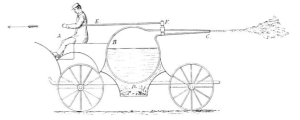

Sir Isaac Newton's steam locomotive of 1680. The fire (D) raises steam in the boiler (B); the driver (A) opens the cock (F), steam issues from nozzle (C) and the result is either incredible excitement or total disappointment. It is unlikely that such a locomotive was built.

Cugnot's steam carriage – inefficient, slow and cumbersome, but a 'first'.

Cugnot's steam carriage – the enormous mass to be turned with the front wheel is clearly visible here.

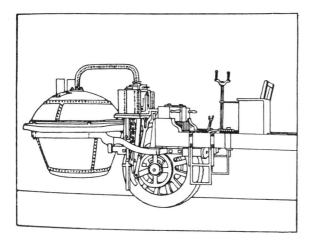

Sir Isaac Newton (1642–1727) has been credited with inventing a vehicle propelled by the reaction of an escaping jet of steam (1680). This assertion is probably an error occasioned by someone suggesting a steam-propelled carriage as an example of the application of Newton's third law: 'For every action, there is an equal and opposite reaction' (often rendered as: 'action and reaction are equal and opposite').

In 1730, a Scots inventor proposed propelling ships by firing guns from the stern. The invention was rejected after experiment, when it was found that three barrels of gunpowder (possibly 300 kg or more) were needed to move the ship for a mile (1.6 km) – 'and still', wrote a contemporary 'there are people who complain that government departments stifle invention.' Little has changed.

Other inventors sought to apply steam to models rather than to full size vehicles. Early steam engines were not suited to driving vehicles. They were slow, cumbersome and inefficient; moreover they produced reciprocating (back and forth) rather than rotative (round and round) motion until crank and flywheel were added. They approached a more suitable form in the later 1700s, and the first working machine was built by the Frenchman Nicolas Joseph Cugnot (1725–1804) in 1770–71 at the Royal Arsenal in Paris by order of the Duc de Choiseul, Minister of War, for dragging artillery on to the battlefield. Cugnot's machine travelled for 12 or 15 minutes at about 2.25 miles an hour (3.6 kph) before it had to stop to generate a fresh supply of steam.

By the time Cugnot had built his second machine, the Duc was in exile and there was no one in power sufficiently interested to run it. The story is that it was tried and that it knocked down a wall, but later examination is said to have revealed that the grate and boiler had never been used and bore no marks of having knocked down anything.

Cugnot's boiler is roughly spherical with an internal fireplace and two flues terminating in short chimneys. It is supported on the fore carriage – which steers the machine – by iron brackets. The engine has two single-acting inverted cylinders which work the front wheel by means of pawls and ratchets, rather than cranks. The workmanship and finish of the machine were particularly good for the time but, with hindsight, the principle is extremely crude.

William Murdock (1754–1839), who was the agent in Cornwall for the steam engine firm of Boulton and Watt at the time, built a model steam carriage in 1784. Murdock is said to have frightened the local parson by running his model on the path leading to Redruth church one night. However, the parson's daughter recalled that her parents were startled rather than frightened, and that Murdock – perhaps because he knew this his experiments would be frowned upon by his employers – asked them to keep his activities secret. But to no avail: Murdock was recalled to Boulton and Watt's Soho Works in Birmingham when Boulton happened across him travelling to London to patent his steam carriage. The fact that he was not dismissed on the spot shows the esteem in which his employers must have held him. Whatever happened, Murdock was not destined to become the pioneer of the steam carriage.

At about the same time as Murdock was building his model steam carriage in Cornwall, William Symington

Murdock's model steam carriage of 1784.

The line-drawing below shows how the long lever transmits the power from the piston to the crank on the rear axle.

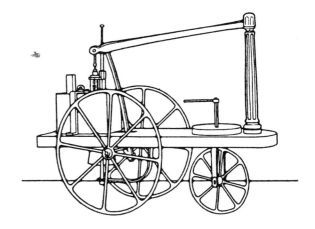

(1763–1831) was building one in the wilds of Lanarkshire in Scotland. Symington was the engineer of the Wanlockhead Mines, and is best known in connection with the early history of steam navigation. He showed his model at Edinburgh in 1786; it worked very well but he was deterred from proceeding further because of the state of the roads in Scotland and the difficulty of obtaining supplies of fuel and water. He therefore turned his attention to designing steam boats, with greater success.

Although Oliver Evans (1755–1819), the celebrated

American engineer and millwright, never built a steam carriage, he was a firm believer in their practicability. In the course of his career he built a number of steam engines, but his many attempts to persuade capitalists to invest in steam carriage projects were fruitless. At last, in 1804, he showed that his plan was not entirely visionary by modifying the steam dredger, which he had built for maintaining Philadelphia harbour, so that it could propel itself from his works to the waterside, a distance of 1.5 miles (2.4 km). Evans wrote:

This was a fine opportunity to show the public that my [steam] engine could propel both land and water carriages and I resolved to do it. When the work was finished, I put wheels under it; and then it was equal in weight to 200 barrels of flour, and the wheels were fixed on wooden axle trees for this temporary purpose, and in a very rough manner, and attended with great friction of course, yet, with this small engine, I transported my great burthen to the Schuylkill with ease; and, when it was launched into the water, I fixed a paddle wheel at the stern, and drove it down the Schuylkill to the Delaware, and up the Delaware to the city, leaving all the vessels going up behind me at least half way, the wind being ahead.

Another American inventor, Nathan Reid, devised a steam carriage in about 1788. He applied for a patent but the application fell through. He proposed a multi-tubular boiler – an important step in the development of the steam carriage, for without some such construction it is impossible to generate sufficient steam within the limits of weight practicable in such vehicles. Reid proposed two double-acting engines each connected to one of the rear wheels by racks and a pinion. This would be the means by which the carriage could turn. The separate engines, powering independent wheels, were designed to enable the carriage to turn – the same principle adopted in present-day tanks. However, his idea of providing additional power by discharging the exhaust steam to the rear (shades of Sir Isaac Newton) would have added little.

Having constructed several successful models another Cornishman – like Murdock, living in Redruth – Richard Trevithick (1771–1833), decided to build a full-sized carriage. He finished it toward the end of 1801 and ran a trial trip on Christmas Eve vividly described by a resident of Camborne:

The year 1801 on Christmas Eve, towards evening Captain Dick [Trevithick] got up steam out in the high road, just outside the shop at the Wieth. When we seed that Captain Dick was a-going to turn on the steam, we jumped up, as many as could – maybe seven or eight of us. T'was a steepish hill going up from the Wieth up to Camborne beacon but she went up like a little bird. When she had gone about a quarter of a mile, there was a roughish piece of road covered with loose stones. She didn't go quite so fast and as it was a flood of rain, and we were going very squeezed together, I jumped off. She was going faster than I could walk. Her went on up the hill about a quarter or half a mile further, when they turned her and came back to the shop.

Trevithick went on to produce a locomotive to run on rails, thus leaving our story. However, pioneers still persisted with steam road vehicles, some of which we will look at in the next chapter.

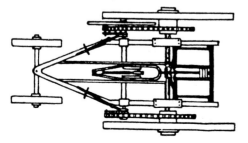

Trevithick's steam carriage of 1802.
The driving cylinder is mounted horizontally, and the reciprocating motion of the piston rod drives a crank on a shaft with a spur gear on each end. These gears mesh with other gears on the back wheels to transmit the drive. Getting in and out of the coach must have been tricky.

Steam omnibuses

Steam omnibuses

In the 1820s and 1830s in England, many people started to experiment with, build and run road-going steam vehicles. The upsurge of activity was such that in 1831, a Select Committee of the House of Commons was appointed to enquire into and report upon the tolls and prospects of land carriage by means of wheeled vehicles propelled by steam upon common roads. The Committee believed 'that the substitution of inanimate for animal power, in draught on common roads, is one of the most important improvements in the means of internal communication ever introduced.'

However, the Committee realised that 'one very formidable obstacle will arise from the prejudices which always beset a new invention, especially one which will at first appear detrimental to the interests of so many individuals.'

Tolls to an amount which would utterly prohibit the introduction of steam-carriages have been imposed on some roads; on others, the trustees have adopted modes of apportioning the charge, which would be found, if not absolutely prohibitory, at least to place such carriages in a very unfair position as compared with ordinary coaches.

The report then summarized the evidence that had been brought to it by the various coach operators, some of whom we will meet shortly, and concluded:

'The Trustees of the Liverpool & Prescot road have already obtained the sanction of the legislature to charge the monstrous toll of 1s 6d per "horse-power", as if it were a national object to prevent the possibility of such engines being used.

'Sufficient evidence has been adduced to convince your Committee:

1 That carriages can be propelled by steam on common roads at an average rate of ten miles per hour.

2 That at this rate they have conveyed upwards of 14 passengers.

3 That their weight, including engine, fuel, water, and attendants, may be under three tons.

4 That they can ascend and descend hills of considerable inclination with facility and safety.

5 That they are perfectly safe for passengers.

6 That they are not (or need not be, if properly constructed) nuisances to the public.

7 That they will become a speedier and cheaper mode of conveyance than carriages drawn by horses.

8 That, as they admit of greater breadth of tyre than other carriages, and as the roads are not acted on so injuriously as by the feet of horses in common draught, such carriages will cause less wear of roads than coaches drawn by horses.

9 That rates of toll have been imposed on steam carriages which would prohibit their being used on several lines of road, were such charges were permitted to remain unaltered.'

An excellent summary of the state of play; as the Committee reported, there were many steam vehicles running satisfactorily, even upon the poor roads of those days. Clearly, the desire to improve this form of transport, if powerful enough, could have brought about a sustained programme of road improvement. However, the money and effort started to pour into the railways instead.

One reason lay in the somewhat diffuse nature of expenditure on roads: how would the money be raised, how would it be used, and how would its use be protected? If the people had wanted a decent road system to carry goods and themselves about the country, it could have been set up years earlier; as it was, the system of levying tolls to maintain turnpike roads hardly resulted in an ideal network with vehicles regularly plying between towns and cities and stopping at villages on the way.

The advantages of the railway over the road were twofold. First, they were run by private companies over land owned by the company, which could therefore control the traffic and levy tolls to make a profit and extend the system. Second, less power was needed to run locomotives and rolling stock over smooth rails laid in as level a manner as possible. What shrewd investor would put money into roads rather than railways?

One further point should be made. The railways, which had been born in mining and quarrying districts, were first seen as a means of moving freight. It was something of a surprise to the proprietors of the Stockton & Darlington railway when they found that people wanted to travel between the two termini, and they had hurriedly to produce a railway coach. As the railways spread, the virtues of mobility were sung and the network grew, bearing the seeds of its own downfall as branch lines were built to serve practically every village and hamlet in the land. Meanwhile, the expectation of door-to-door travel, which had been set up by the railways, but which they could never meet, gradually brought into being a road system which eventually eroded their advantages more and more.

One commentator wrote: 'The arguments levelled against the road carriage as to the ill effects which would result from the displacement of a large part of the 2,000,000 of horses then in use for transport uses, were powerless against the railway locomotive running under statutory powers.' The steam coach, therefore, which had reached a high state of development between the years 1825 and 1832, came at an unfortunate time, and died of lack of interest.

Early steam coaches

The steam coach movement seems to have started with Julius Griffiths of Brompton who, in 1821–22, had a carriage built by Joseph Bramah, the well-known engineer, sanitary plumber and locksmith. The advent of this carriage was announced with a great flourish of trumpets. Because of its 'universal importance' it was patented in England, Austria and America. It was supposed to carry three tons of merchandise at 5mph (8 kph) at a saving of 25% compared with horses. But in spite of the best workmanship of the day, it was a complete failure.

It had a very long wheelbase and the engine and boiler were hung over the rear axle by means of a spring suspension. The steam was generated in a water-tube boiler, made up of a number of rows of horizontal tubes which extended across the fire box and passed out through its sides into the vertical connecting boxes. The two-cylinder engine had a condenser consisting of a number of flattened tubes of thin metal cooled by air. The boiler seems to have been the greatest source of trouble, for it was found impossible to retain the water in the lowest tubes.

David Gordon worked away at his steam-carriage project at about the same time as Griffiths – and with no greater success. One idea was to place the engine inside a squirrel cage and drum nine feet (2.74 m) in diameter. The wheels of the engine were toothed and the idea was that it would try to climb up inside the drum, thus pushing it forward. The engine pushed a carriage in front of it by means of side rods.

Gordon's scheme patented in 1824 was to have legs, or 'propellers' to push it along. It was steered by raising the legs on one side or the other.

About the year 1825 Henry Peto built a steam coach propelled by a beam engine and provided with a two-speed gear. Not surprisingly it was found too unwieldy and heavy for road traffic, and never got beyond the experimental stage. In the same year T W Parker in America is said to have made a large model

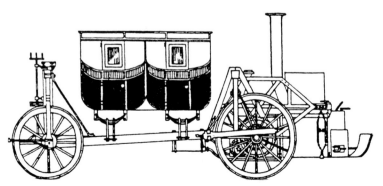

Julius Griffiths's steam carriage of 1822, one of the first to take to the road. The rear wheels are driven by spur gears. The carriage in the picture travels to the left.

David Gordon's ingenious solution combining rail and road: the rails on which the locomotive runs are formed by two rings nine feet in diameter. The large diameter of the rings enables the carriage to traverse rough ground. Unfortunately, the funnel of the locomotive sticks up between the rings, so there cannot be cross-members linking them – most unstable!

David Gordon's steam carriage of 1824, designed to push itself along with reciprocating legs. Apart from the fact that it moves to the left, it looks more like our idea of a railway locomotive than did railway locomotives of the time.

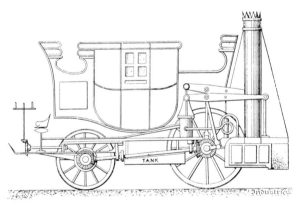

Burstall & Hill's steam coach of 1824 has two pistons driving cranks on the rear axle. The axles are connected by a longitudinal shaft with gears at each end. The seats for driver and passengers give some idea of the scale of the machine.

steam carriage mounted on three wheels eight feet (2.44 m) in diameter propelled by a two-cylinder steam engine. James Nasmyth, of steam hammer fame, built a satisfactory model which he turned into a full-size steam carriage about 1827. It worked very well, carrying eight passengers but after being used experimentally for a few months was broken up for no apparent reason.

Walter Hancock (1799–1852)

Hancock was one of the pioneers who brought the steam coach to a high state of development. He built a number of coaches including *Autopsy*, *Enterprise* and *Era*, all of which ran with powerful engines on high-pressure steam. His *Automaton* (1836) carried its passengers and conductor and driver on seats in the front part of the vehicle set in the form of a char-à-banc – literally, a carriage with benches. The engine and boiler were mounted in the rear part of vehicle with space for some 'inside' seats between them (which must have been somewhat hot and uncomfortable). The vertical engine was set about half way along the vehicle; it had two cylinders of nine inches (22.9 cm) bore and 12

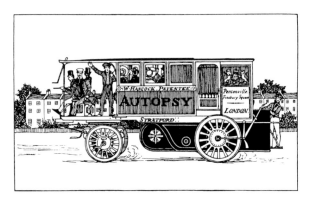

Three of Walter Hancock's steam coaches; *Autopsy* and *Era* of 1833, and *Automaton* of 1836. Hancock paid much attention to the comfort of his passengers, but lack of official recognition of his contribution to civilisation (and the increasing road tolls) drove him to pursue an interest in railways.
Autopsy ran daily for 24 weeks for hire between Finsbury Square and Pentonville in north London. *Automaton* ran about 4,200 miles (6,800 km) in 20 weeks, carrying 12,700 passengers between the City of London and Paddington in the west, Islington in the north, and Stratford in the east.

inches (30.5 cm) stroke and a crankshaft geared to the rear driving axle by means of a chain. The engine was placed in a separate compartment so that it could be properly cleaned and maintained, and for the first time a clutch was fitted so that when the carriage was stationary the engine could still run to drive the boiler feedwater pump and furnace fan. The driving wheels were 4ft (1.22 m) in diameter and the coach weighed some 3.5 tons.

Hancock's omnibus ran at 10 mph (16 kph), at which speed the engine turned at 70 rpm. The boiler ran at a pressure from 60 to 100 lbs per square inch (0.4 – 0.7 MPa), though he had experimented with pressures as high as 400 psi (2.76 MPa). Hancock's boiler had ten chambers, each about 30 x 20 x 2 in (76.2 x 50.8 x 5.1 cm) in thickness arranged so that the hot products of combustion could pass through. The construction was admirably suited to raising a sufficient quantity of high-pressure steam to run the vehicle.

Sir Goldsworthy Gurney (1793–1875)

Gurney was yet another Cornishman, who had watched Richard Trevithick working with his models. He was born at Padstow and trained as a doctor, setting up in practice as a surgeon at Wadebridge where he married in 1814. Six years later he moved to London, where he practised as a surgeon, delivered lectures on chemistry *etc*, and was soon moving in the highest scientific circles of the day. He invented many useful things, including the high temperature flame produced by burning oxygen and hydrogen, the brilliant 'limelight' used in the theatre, and the steam jet for ventilating mines. In 1852 he arranged the lighting and ventilating systems of the new Houses of Parliament which he continued to superintend until 1863, when he was knighted. He died in 1875.

Gurney began to experiment with steam carriages in 1823 and after one or two experiments brought out his 'new improved steam coach' in 1827. It carried 21 passengers, and had a water-tube boiler to raise the steam, and propelling legs to help the carriage move should the wheels slip round and round. He soon found, however, that the legs were not needed.

The key to the diagram (*opposite*) is as follows:

1 The Guide and Engineer, to whom the whole management of the machinery and conduct of the Carriage is entrusted. Besides this man, a Guard will be employed.

Sir Goldsworthy Gurney's steam coach in cross-section. The horizontal engine drives a crank on the rear axle and the eccentric (P) operates the slide valve. Steering with the somewhat spindly handle (X) must have taken some effort.

Gurney's first steam coach, 1827.

2 The Handle, which guides the Pole and Pilot Wheels.

3 The Pilot Wheels.

4 The Pole.

5 The fore Boot for luggage.

6 The 'Throttle Valve' of the main steam-pipe, which, by means of the handle, is opened or closed at pleasure, the power of the steam and the progress of the carriage being thereby regulated from 1 to 10 or 20 miles an hour.

7 The Tank for water, running from end to end and the full breadth of the carriage: it will contain 60 gallons of water.

8 The Carriage, capable of holding six inside passengers.

9 Outside Passengers, of which the present carriage will carry 15.

10 The Hind Boot, containing the Boiler and Furnace. The Boiler is encased in sheet-iron, and between the pipes the coke and charcoal are put, the front being closed in the ordinary way by an iron door. The pipes extend from the cylindrical reservoirs of water at the bottom to the cylindrical chamber for steam at the top, forming a succession of lines something like a horse-shoe turned edgeways. The steam enters the 'separators' through large pipes, which are observable on the Plan, and is thence conducted to its proper destination.

11 'Separators,' in which the steam is separated from the water, the water descending and returning to the boiler, while the steam ascends and is forced into the steam-pipes or main arteries of the machine.

12 The Pump, by which the water is pumped from the tank, by means of a flexible hose, to the reservoir communicating with the boiler.

13 The Main Steam-Pipe, descending from the 'Separators', and proceeding in a direct line under the body of the coach to the 'throttle valve' (No 6), and thence under the tank to the cylinders from which the pistons work.

14 Flues of the Furnace, from which there is no smoke, coke and charcoal being used.

Sir Charles Dance's steam coach of 1833. The traction locomotive is separate from the passenger vehicle – a forerunner of the later traction engine.

15 The Perches, of which there are three, conjoined, to support the machinery.

16 The Cylinder. There is one between each perch.

17 Valve Motion, admitting steam alternately to each side of the pistons.

18 Cranks, operating on the axle; at the end of the axle are crotches (No 21) which, as the axle turns round, catch projecting pieces of iron on the boxes of the wheels and give them the rotary motion. The hind wheels only are thus operated on.

19 Propellors, which, as the carriage ascends a hill, are set in motion and move like the hind legs of a horse, catching the ground and thus forcing the machine forward, increasing the rapidity of its motion and assisting the steam power.

20 The Drag, which is applied to increase the friction on the wheel in going down a hill. This is also assisted by diminishing the pressure of the steam – or, if necessary, inverting the motion of the wheels.

21 The Clutch, by which the wheel is sent round.

22 The Safety Valve, which regulates the proper pressure of the steam in the pipe.

23 The Orifice for filling the tank. This is done by means of a flexible hose and a funnel, and occupies but a few seconds.

Like many pioneers, Gurney had to put up with stupid opposition as well as downright hostility. On one occasion, he made the journey to Bath with a number of guests in his carriage. They averaged 12 mph (19.3 kph) until they reached Melksham, where there was a fair. It was impossible to drive through the dense crowds and, since they were mainly agricultural labourers and manufacturers of the district, and considered all machinery injurious to their interests, they set upon the carriage and its occupants chanting: 'Down with all machinery'. A passenger, Mr Bailey, Gurney and his assistant engineer Thomas Martin were seriously injured, and were taken unconscious to Bath in another carriage.

Gurney was not put off and continued to build carriages. He supplied them in 1831 to Sir Charles Dance, who started a service between Cheltenham and Gloucester. Local feeling still ran high, but the project was successful and made a profit over some 400 journeys. The service continued until 1840 when it – and all similar enterprises – were killed by the imposition of ruinously heavy tolls on self-propelled road vehicles. Gurney's price for a steam coach was £1,000 and his royalty on all public services using the coach was sixpence a mile. His losses caused by the imposition of the tolls were so heavy that a committee in the House of Commons appointed to consider the matter recommended that he should receive a grant of £16,000 in recognition of his public services.

There were many other steam carriages, some of which are illustrated here. Between 1832 and 1838, as the railways were coming into their own, a dozen carriage companies worked in various parts of the country, and there was no shortage of engineers to design and build the vehicles. But the tolls levied on their operation were prohibitive – for example:

The patent steam carriage of Sir James Anderson, Bart and W H James Esq. How the helmsman controls the vehicle, and what room there is for the passengers, is unclear.

Dr Church's steam coach (1833) was indeed a marvellous construction, in outline and ornamentation something between a gypsy van, a merry-go-round, and a ship's saloon. It must have looked – and sounded – imposing, if not terrifying.

On the Liverpool and Prescot road Mr Gurney would be charged £2.8s, while a loaded stage coach would pay under 4/-. On the Bathgate road the same carriage would be charged £1 7s 1d while a coach drawn by four horses would pay 5/-. On the Ashburton and Totnes road [in Devon] Mr Gurney would have to pay £2, while a coach drawn by four horses would be charged only 3/-.

Gradually, the steam coach disappeared from the roads and road haulage by steam was focused on the traction engine drawing a trailer or trailers. As these machines developed, the toll laws were reassessed and by 1865 there was a uniform scale of tolls throughout the country. At the same time, the Acts regulated the weight and speed of the vehicles in such a way as to kill passenger carriage and drive auto-locomotion off the roads. According to one sore commentator:

All the pains and expense, all the time and patience, that had been devoted to bringing this industry to the point at which it had arrived in 1840 were, if not actually wasted, at any rate robbed of their due development and reward. The industry

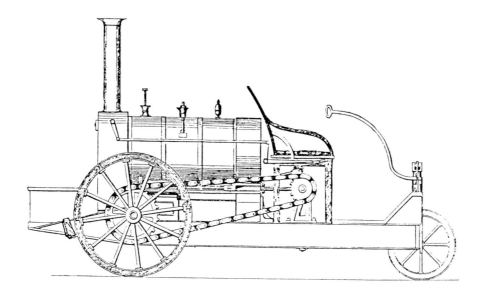

Rickett's steam carriage of 1858. This vehicle is more like a private car than an omnibus.

Rickett's improved steam carriage of 1861 had a final drive using gears rather than the prominent chain of 1858.

Garrett's steam carriage of 1861 weighed five tons and seated nine people – in addition to the staff.

The driver's position on the ABC single-cylinder steam velocipede (1869) seems to need some development. The steering wheel could be easier to reach – the coal shovel more so.

in which England should have led the world was left to be taken up by other nations, who are still reaping the profit of their thirty years' start.

The increased tolls ran early steam vehicles off the roads by 1840, but the railway still had one shortcoming – that it did not provide door-to-door transport. One still had to get to the station and from the station, however marvellous the journey between, steam whisking one effortlessly from a station at one end of the country to another station at the other. (Later manufacturers would find that, if they had to load a vehicle to take goods to the station, they might as well go straight to their destination.)

It was this shortcoming that kept a few pioneers at work; one was Thomas Rickett of Buckingham, who began building steam carriages in 1858. His first was like a three-wheeled steam locomotive, the driver sitting in front of the boiler, with the cylinders beneath the seat driving the road wheels by means of a chain. The single front wheel had tiller steering. Two passengers could share the driver's bench seat, while a stoker stood on a platform at the back. Normally, drive was to one wheel only, but there was a clutch for connecting the wheels when road conditions demanded. The 1.5

ton machine reached a maximum speed of 12 mph (19.3 kph).

Rickett built an improved vehicle in 1860 which drove the wheels via spur gears rather than via a chain. In his third and final vehicle, the pistons drove the wheels via cranks, as in a railway locomotive.

Describing the steam coaches, an observer recalled that they were

Vast, unshapely bodies perched on uncouth frames and monstrous wheels, weirdly decorated, childishly emblazoned, riotously extravagant in bulk and weight, top-heavy and ill-balanced, grotesque and formidable, terrifying and ludicrous, belching clouds of black smoke and showers of cinders, enveloped in dust, their passage accompanied by the shrieks and barks of dogs, the whinnying panic of horses, the terror and delight of children, the wonder and admiration of the polite world – are they not, after all, typical of that joyful, sanguine enthusiasm over a new discovery that can hope all things, admire all things, endure all things? And would not one like to have been there some sunny morning when the *Autopsy* was starting from the Angel towards the Euston Road, or when the *Era*, turning the corner of some Wiltshire greenwood, passed triumphantly on its way from London to Marlborough, with the children cheering and the dogs barking? We may be sure that its passengers tasted and enjoyed to the full those little pleasures and excitements of the open road – pleasures and excitements that we have only lately rediscovered.

The power of steam had so far been the main driving force for self-propelled road vehicle experiments. Vehicles driven by huge coil springs were ruled out because of the mechanical problems and the enormous weight of material necessary to store the power. Compressed air was tried, but a large part of the weight of the vehicle lay in the stout vessel needed to store air at a pressure sufficiently high to give it a satisfactory working range. An attempt was made to apply the principles of the atmospheric railway, by attaching the vehicle to a piston running in a tube and exhausting the air at the front to push the piston along. However, without the rails, the arrangement was even less satisfactory. Electricity was an up-and-coming source of power and interest but, as with compressed air, the mass of the storage medium – in this case the cells to generate the electricity – was prohibitive. The world was waiting for the internal combustion engine.

ENGLISH MECHANIC
AND MIRROR OF SCIENCE
Engineering, Building, Inventions, Electricity, Photography, Chemistry, &c.

VOL. IX.—No. 214.　　　FRIDAY, APRIL 30, 1869.　　　[PRICE TWOPENCE.

Caleb Williams thoughtfully furnished his locomotive with a roof attached to the funnel to prevent condensed steam and soot falling on to the passenger; it also acts as a sunshade. The picture is an artist's impression of 1869; Williams had built only a model.

Left. Cowan's improved steam carriage for use on common roads (1869) follows railway locomotive practice with the pistons driving outside cranks on the road wheels. Once again, the arduous job of stoker is underplayed.

Charles Randolph of Glasgow built a steam carriage in 1872. It was 15 feet (4.5 m) long, weighed 4–5 tons, travelled at 6 mph (9.6 kpm) and could accommodate the driver and two passengers in the front box and six more passengers in the body.

The engine

The idea of the internal combustion engine is even older than that of the piston steam engine. In 1680, the Dutchman Christiaan Huygens (1629–1695) built an engine in which a weight was raised by the explosion of a mixture of gunpowder and air in a closed system fitted with a piston.

When the piston is at the bottom of the cylinder, gunpowder is exploded inside. This drives the piston up; then the products of combustion are driven out and a partial vacuum forms inside the cylinder. Atmospheric pressure forces the piston down into the cylinder, raising a weight. The process must have been slow and cumbersome – not to mention dangerous. Huygens therefore considered using the elastic power of steam, but died before he could build and test a model.

Huygens' experiments were continued by the French physicist (and inventor of the pressure cooker) Denis Papin (1647–1714), and the principle led, first to Savery's 'fire-engine', and then to Newcomen's steam engine in which condensing steam in the cylinder causes the piston to be pushed down by atmospheric pressure.

Newcomen's engine came into widespread use for pumping water from mines and formed the inspiration for James Watt (1736–1819) who increased its efficiency by condensing the steam outside the cylinder, and continued to develop many improvements, including the double-acting steam engine (steam admitted alternately to one side of the piston and the other) and the crank, providing rotative power for driving machinery.

As the 19th century dawned, the extraction of coal gas, and its use for lighting and heating, began to develop. In due course, the explosive properties of a gas–air mixture were exploited for the internal combustion engine – that is, one in which the combustion takes place inside the cylinder, doing work in pushing a piston along.

By the mid-19th century, several builders were experimenting with internal combustion engines burning gas; the first to succeed was the Frenchman Etienne Lenoir (1822–1900) in 1859. Lenoir's first engine was a converted steam engine. It had a horizontal cylinder with a crosshead and flywheel and Lenoir used the slide valve gear to admit the gas and to allow the products of combustion to exhaust. It was double-acting (that is to say, the forces of combustion acted first on one side of the piston; then on the other) and the gas–air mixture was ignited by two sparking plugs, one in each end of the cylinder. The motion of the piston pushed out the exhaust gases from the other side of the cylinder.

Adopting steam-engine practice inhibited the discovery of the principle of compressing the charge of gas, so that the engine was somewhat inefficient, using perhaps ten times as much fuel as a modern engine would. As can be imagined, Lenoir's engine ran at a very much higher temperature than the steam engine on which it was based. Although it was water cooled, it suffered from overheating and from inadequate lubrication.

In 1862, M Hugon, a French engineer, built an engine on Lenoir's principles, in which a fine spray of water injected into the cylinder after the explosion helped cooling. This reduced the gas consumption and lowered the temperature of the exhaust gases in comparison with the Lenoir engine, but it was still not very satisfactory.

Building gas engines on steam engine practice was not the way forward. Lenoir continued to work on the gas engine and in the mid-1880s brought out a single-acting compression engine which used about a quarter of the amount of gas consumed by his earlier ones.

In 1862 another Frenchman, Alphonse Beau de Rochas (1815–91), patented the 'four-stroke cycle'; a 'stroke' being the travel of the piston from one extemity to the other. In a vertical engine, on the first stroke of the cycle the descending piston draws the mixture into the cylinder, on the second stroke it

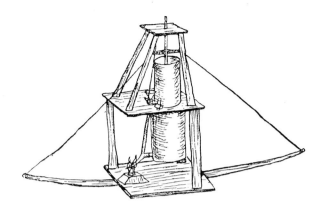

The aviation pioneer Sir George Cayley sketched his Gunpowder Explosion Engine on 22 November 1807. When running the engine itself opens a cock below the little funnel, introducing a small charge of gunpowder into the pipe heated by flames (left of base plate). The expanding gases push up a piston, which stretches the 'bow string', bending the ends of the 'bow' at the base of the machine upwards. At the top of its stroke, the piston opens a stop cock, releasing the gas inside and allowing the bow to pull the piston down again, 'the power of the stroke being applied to any work proposed'.

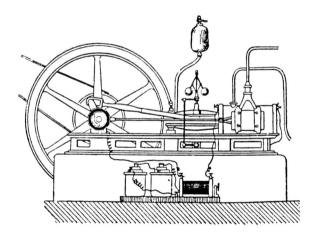

Lenoir's first engine (1860). In this adapted double-acting steam engine, gas is substituted for steam, and a spark plug at each end ignites the gas, causing it to expand, pushing the piston across, which drives out the exhaust gases from the previous ignition. The batteries and trembler coil to produce the spark – one side connected to the plugs and the other to an earthed distributor – are seen at the base.

compresses it, the third stroke begins with ignition which drives the piston down, and the fourth stroke pushes the spent gases out. Beau de Rochas did not exploit his invention and allowed his patent to lapse. However, it was he who laid down the principle that the internal combustion engine should run at high speed, and he who showed the importance of compressing the explosive mixture before igniting.

Otto and Langen's free-piston engine

In 1866, the German engineers Otto and Langen built a 'free piston' vertical engine in which the piston was connected to the power shaft not by a crank, but by a rack and pinion with a free wheel. Here, the charge was admitted to the bottom of the cylinder and, when it was exploded, it drove the piston up. When the gases had expanded and cooled, the piston descended by the force of gravity and the rack and pinion pulled the fly wheel round. At the same time an exhaust valve opened so as to let the spent gases out. At the bottom of its stroke the piston was raised slightly, and a fresh charge of gas and air was admitted and ignited when a shutter opened and exposed a gas flame. (The explosion would blow out the gas flame, but there was another flame ready to re-ignite it when the shutter closed!) Inefficient and noisy as it was, the Otto and Langen engine worked more reliably than one might imagine. It consumed about a third of the gas of Lenoir's earlier engines, and enjoyed some popularity in its day.

The gas engine was a valuable substitute for the steam engine when gas was available at a reasonable price. The idea was emerging that liquid hydrocarbons (compounds of carbon and hydrogen which are the chief constituents of natural gas and crude oil) might be vaporised and used to form an explosive mixture with air – though their availability was limited owing to lack of demand at that time. In 1873 J Hock of Vienna patented an engine in which air under pressure broke up a jet of oil into a fine spray, which was then injected into the cylinder. In the same year, Brayton of Philadelphia invented an engine with two cylinders, one for compression and the other for working. Air was forced through absorbent materials soaked in hydrocarbons to become charged with vapour. The mixture was admitted to the working cylinder and fired, driving the piston down; the returning piston expelled the

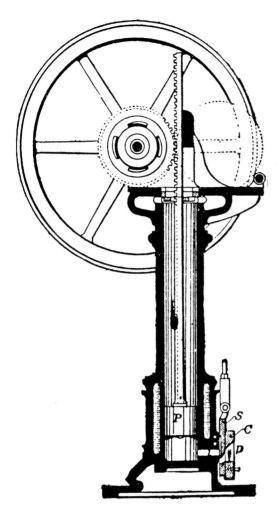

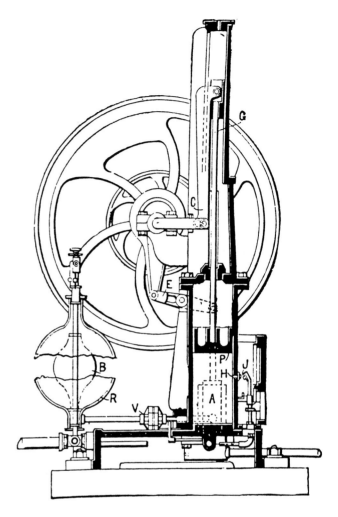

Free piston Otto & Langen engine. When the flame at D is unshuttered, it ignites the gas mixture below piston P, driving it to the top of its travel. The gases cool and piston P falls by gravity; the freewheel is locked and the flywheel pulled round.

The Bisschop two-stroke engine. Gas and air are drawn into the lower end of the cylinder P and through inlet valves V for about one third of the upstroke. A small flap valve H opens inwards, allowing a flame J to fire the charge. The piston then completes its upstroke, driving the crank by rod G and connecting rod C. Exhaust piston valve A, driven by eccentric E, opens on the downstroke. The Bisschop engine was noisy, but very simple, and did well in the hands of small-power users with little mechanical knowledge.

A 30-40 BHP Crossley gas engine working on a four-stroke cycle.

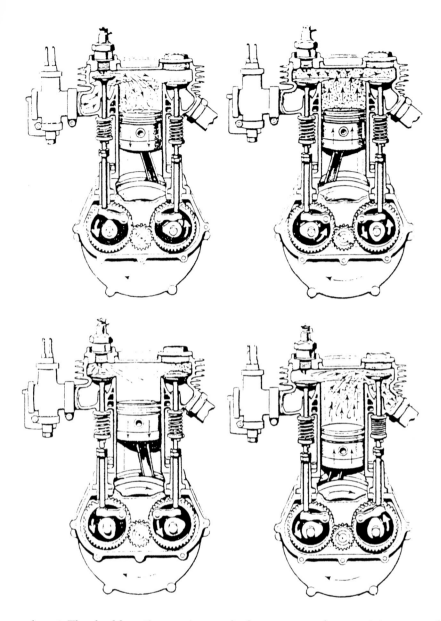

The four-stroke cycle of the internal combustion engine. On the first stroke – induction – the piston descends, the inlet valve opens, and explosive mixture is drawn into the cylinder. As the second stroke – compression – begins, the inlet valve shuts so that the rising piston compresses the mixture. At the top of the stroke the spark passes and the piston is driven down for its third – power – stroke. The exhaust valve opens as the piston rises again for the fourth – exhaust – stroke, pushing the spent gases out.

exhaust. The double-acting engine worked on a two-stroke cycle and was started by compressed air, stored in a reservoir charged when the engine was running.

The Otto cycle

In 1876 Otto produced a new engine, which he called his 'silent engine' – by contrast with his free piston type. Otto based his engine upon the Beau de Rochas cycle, although he had probably never heard of the Frenchman. Since Otto's engine was successful, the foundation upon which most internal combustion engines have been developed ever since – and possibly

because it is easy on the tongue – the four-stroke cycle became known (somewhat unfairly to Beau de Rochas perhaps) as the Otto cycle. The importance of the principle may be judged from the fact that, of 53 engines exhibited at an international exhibition in 1889, all but four used it.

The Diesel engine

Engines with reciprocating pistons working on the four-stroke Otto cycle became the norm for private cars, and many other vehicles. As is well known by anyone who has used a bicycle pump, when a gas is

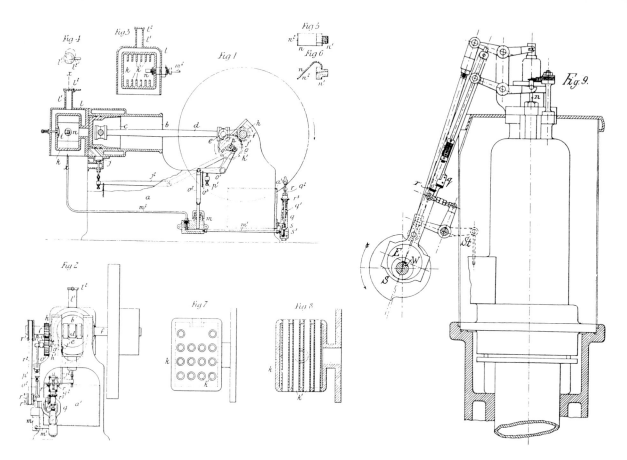

Details from Dr Rudolf Diesel's patent drawings for a compression-ignition engine. When the piston compresses the air in the cylinder, the temperature rises sufficiently to ignite a drop of fuel injected at the right moment.

compressed rapidly it becomes hot. Early engines were prone to overheating because of this effect, as well as from the energy of the combustion itself. Overheating can cause pre-ignition – that is to say, the burning takes place before the piston has reached a position where it can optimise the use of the released power. In an attempt to harness this effect, in 1890, Herbert Akroyd Stuart took out British patent no 7,146 for a compression-ignition engine.

A compression-ignition engine works on the Otto cycle, but the 'compression ratio' (the ratio between the total and unswept volumes of the cylinder) is high. The piston draws air in through an inlet valve as it descends; when it rises again, the air is compressed and becomes hot. Near the top of the stroke, a charge of oil is injected; the heat ignites it to provide power.

Although Stuart's principle was used in engines

produced by the Lincolnshire enginering company Hornsby of Grantham, the compression ratio was not high enough; the engines needed pre-heating before they would run by themselves.

This type of power unit is now known as the Diesel engine, after Dr Rudolf Diesel (1858–1913), a Parisian-born German engineer who covered his principle in British patent no 7,241 of 1892. Diesel understood and explored the thermodynamic principles more thoroughly, and produced an engine which would start from cold. He deserves the credit.

The two-stroke cycle

In the four-stroke cycle there is one power stroke for every two revolutions of the crankshaft. The two-stroke engine has one power stroke for each revolution. In the two-stroke engine, the combustible charge

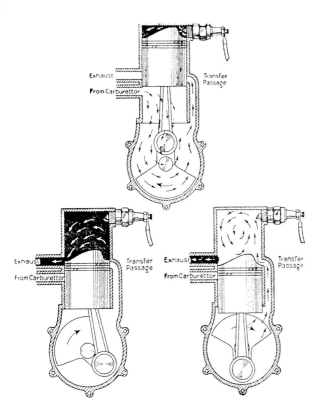

but does round off the story of the internal combustion engine. Clearly, a rotary engine is likely to be more efficient than one which has large masses of metal moving up and down. But it was not until 1957 that Wankel ran an experimental engine in which a triangular-shaped rotor in a shaped chamber accommodated the Otto cycle. The force of the expanding burning gases acts on the surface of the rotor; there are three complete combustion cycles for each revolution. The output shaft runs at three times the rotor speed. The engine is water cooled and has a compression ratio of 8.6:1.

The NSU Spidercar built at Neckarsulm, Germany in 1963 was the first to be powered by a rotary combustion engine. Although the design has its advantages, it was over half a century behind entrenched reciprocating piston engine design and production practice – never an encouragement to novelty.

In the first figure, the plug is sparking, initiating the power stroke. At the same time, the combustible mixture is flowing into the crankcase. In the second figure, the spent gases are escaping from the exhaust port uncovered by the piston. In the third figure, the piston reaches the bottom of its stroke and uncovers the transfer port; the compressed mixture flows from the crankcase and, because of the shape of the piston head, helps to push out the vestiges of spent gases.

is drawn into the crank case and swept up to the combustion chamber through a transfer port as the exhaust gases from the previous charge are swept out of an exhaust port. The shapes of all the components involved are clearly crucial, and the satisfactory development of the two-stroke engine (used mainly for motor cycles, lawn mowers and so on) has depended on proper analysis of gas flows.

The rotary engine

Professor Felix Wankel, later director of the Technische Entwicklung Stelle, started work on his rotary engine in 1954. It is therefore somewhat late for inclusion here,

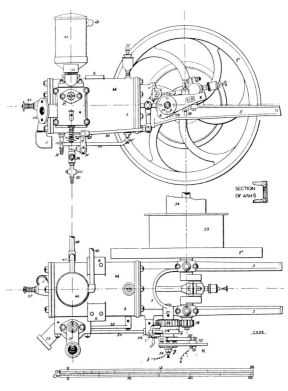

The Benz single-cylinder motor in elevation and plan. The cylinder is to the left, and the flywheel a prominent feature.

Having just passed his driving test in 1951, the author was fortunate enough to spend the afternoons of a week driving a 1907 Darracq round the streets of Cambridge, England to advertise an Accident Prevention Exhibition in the Corn Exchange. In the morning – as seen here – the owner of the Darracq took the wheel and the author became the 'lady' passenger.

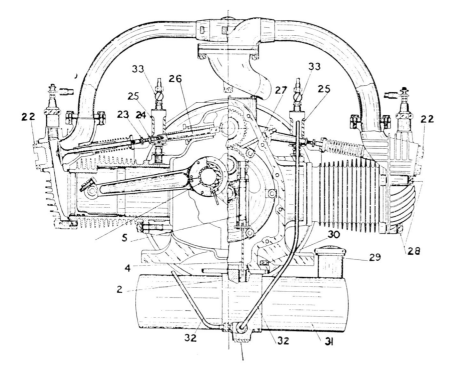

Half-section of the Rover 8 horizontally-opposed twin cylinder engine. The engine was air cooled, and the oil tank at the bottom (31) is arranged to help. The Rover 8 was a very successful British car which succeeded the plethora of light cycle cars which emerged in the teen years of the century. It was itself ousted by the even more successful Austin 7.

The parts of the engine

Given a reciprocating engine working on the Otto Cycle, the scope for basic design variations is somewhat limited. Earlier engines exhausted almost every possibility, starting with the single cylinder engine. In twin-cylinder engines, the cylinders could be in line, Vee, or horizontally opposed. The two-in-line engine with both cranks in the same position had evenly-spaced power strokes, but balancing problems. The two-in-line cranks at 180° had slightly better balancing, but uneven power strokes: chuff-chuff ... chuff-chuff, as in the early Darracq.

The horizontally-opposed twin made the best of all worlds, and was widely adopted – especially for motorcycles, where its compact layout is highly suitable for the vehicle.

Four-cylinder engines became the norm for the mass-produced cars; six- and sometimes eight-cylinder engines for more luxurious models. At this point, the crankshaft tends to lose strength because of its length, hence the development of the V–8 and V–12 engines. One departure to be noted is the Gobron-Brillié opposed piston engine, where combustion takes place between two pistons, supposedly evening out the thrust on the crankshaft.

The valves

Having set up the cylinder and crankcase with pistons and crankshaft, one needs to provide some means of introducing the combustible mixture, and removing spent gases. Early practice used the slide valve of the steam engine, but this was for expediency only. Poppet valves became the norm, kept shut by means of springs on their stems, and opened by means of cams.

The Daimler two-cylinder V-engine. The circular dotted lines in the right-hand view represent a track in which followers run to open the valves once every two revolutions.

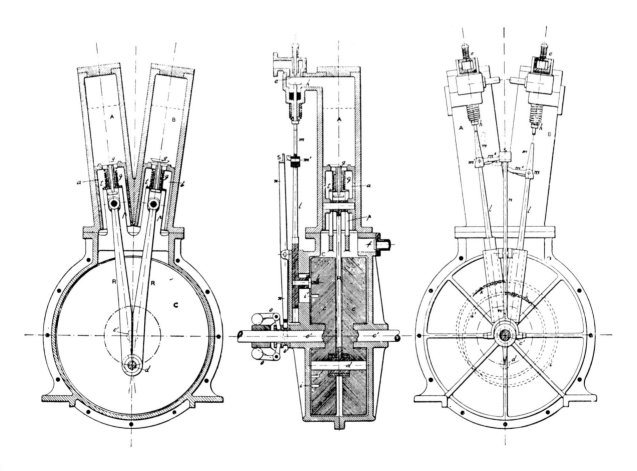

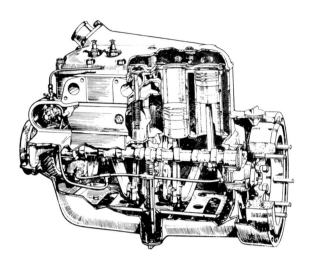

The fearsome Gobron-Brillié engine with the top cover removed.

The Morris Cowley four-cylinder engine partly sectioned.

Diagram of Gobron-Brillié engine, an opposed engine turned inside out, thus saving one set of spark plugs and valves. Such a saving was dwarfed by the problems thus introduced.

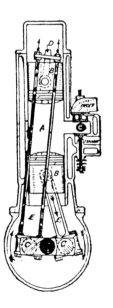

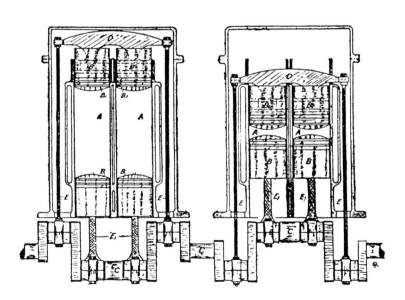

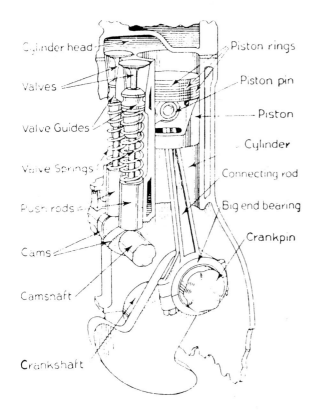

Cylinder head

Valves

Valve Guides

Valve Springs

Push rods

Cams

Camshaft

Crankshaft

Piston rings

Piston pin

Piston

Cylinder

Connecting rod

Big end bearing

Crankpin

The moving parts of the four-stroke engine.

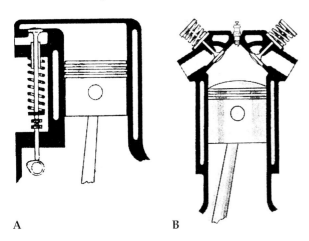

A   B

The poppet valve has a mushroom-shaped head
seated in a conical hole. It is kept shut by a spring,
and opened by a rotating cam. (A) is a side-valve
engine; inlet and exhaust valves are along one side
and intricate passages connect them to inlet and
exhaust manifolds. (B) is an overhead-valve engine,
which can provide a better combustion chamber
shape.

In a four-stroke cycle, the valves operate once every
two revolutions of the crankshaft, so 2:1 gearing is
used to drive the camshaft. Sometimes the inlet valve
was 'automatic' – that is to say reduced pressure on the
induction stroke caused it to open – though a controlled opening is preferable as the rest of the engine design develops.

    The valves are actuated from the camshaft by inter-
mediate tappets, push rods or rockers, according to the
relative positions of the parts. A convenient early valve
layout was overhead inlet (which could be automatic

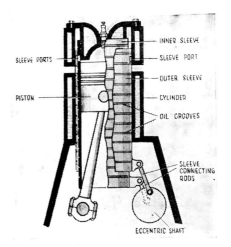

SLEEVE PORTS

PISTON

INNER SLEEVE

SLEEVE PORT

OUTER SLEEVE

CYLINDER

OIL GROOVES

SLEEVE
CONNECTING
RODS

ECCENTRIC SHAFT

The double sleeve-valve engine has two sleeves
sliding along their long axes to uncover the inlet and
exhaust ports as needed.

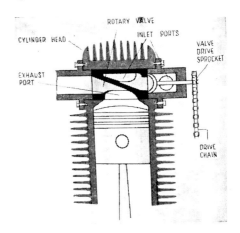

ROTARY VALVE

CYLINDER HEAD

EXHAUST
PORT

INLET PORTS

VALVE
DRIVE
SPROCKET

DRIVE
CHAIN

The rotary valve engine has a rotating element to
provide the same result as poppet valves.

for simplicity) and side exhaust. The reliable, slow-revving engines which became the norm from the turn of the century onwards usually had side valves. Later designs favoured overhead valves because the shape of the combustion chamber could be more regular.

Reverting to steam-engine practice, some engineers designed power units with sleeve-valves, which could provide a silent, though somewhat expensive, solution. Rotary valves were also tried, but the poppet valve remained the most simple and reliable and hence become the most widespread.

Carburation

The mixture of air and combustible vapour (which soon became that of petroleum spirit) is provided by a carburettor. In early versions, the air is drawn over, or is bubbled through, the fuel, or a rotating brush sprays it into the air stream. As engines improved, a better means of controlling the mixture was needed, and designs were based on a constant head of fuel (from a float chamber) supplying a jet (in a venturi tube – shaped so as to increase the air flow past the jet, and hence the vaporizing spray issuing therefrom). Apart from fuel injection systems, the basic principle of the carburettor with jets has remained unchanged.

Other features of the carburettor are the butterfly valve in the inlet pipe, which controls the amount of mixture available, and hence the speed of the engine via the accelerator pedal; and the choke, which cuts down the amount of air in the mixture, making it 'richer' for starting the engine, especially when cold.

Ignition

Apart from the compression-ignition engine, some extra means of igniting the charge is needed. Early attempts included the hot tube, where an incandescent tube causes ignition when the mixture encounters it, and flame ignition (mentioned above) where a shutter allows the mixture to see the flame; the flame however is blown out by the combustion, so there is another flame to re-light it when the shutter closes.

The spark-gap, in the form of the spark plug, came in quite early, and it was soon found that a high-voltage (rather than a low-voltage) spark served, produced either by battery and coil or by magneto. The principle of the induction coil was well understood, and there is a lot to be said for not having to develop too many technologies at once. However, the coil does need a source of electricity, and an answer to that complication was found in the magneto, which generates

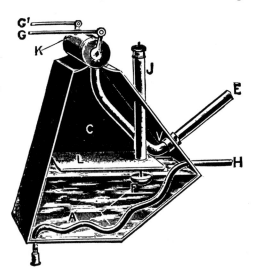

In the early de Dion carburettor, air is drawn over the surface of fuel in a reservoir to provide a combustible mixture.

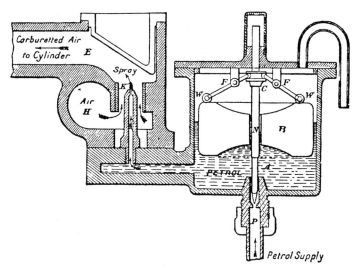

The simple carburettor which became the norm had a float chamber (right) to maintain a constant fuel level, and a spray jet (K); air entrained fuel as it flowed past the jet to the inlet manifold (left). Until the advent of fuel injection, all carburettors were based on this principle.

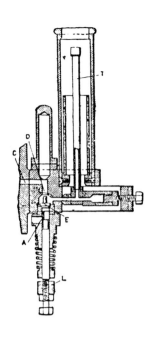

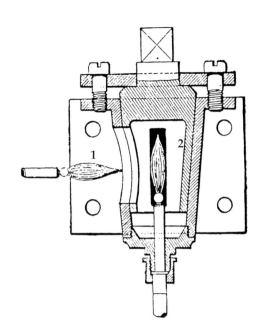

Left

Hot-tube ignition, where the combustible mixture is fired when the tube is uncovered. It was not a very satisfactory method.

Right

Flame ignition – the first flame ignites the mixture when the port opens and the second flame re-ignites the first flame because combustion blows it out.

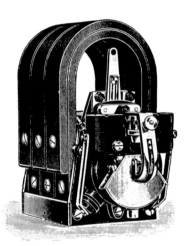

A magneto makes its own electricity; the magnets between whose poles the rotor turns thus generating the pulse of sparking energy are a prominent feature of the four magnetos shown. The contact breaker is housed at the front of the rotor shaft.

Typical spark plugs.

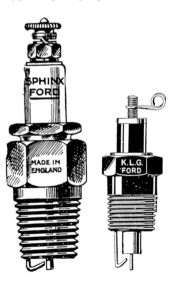

its own electricity and thus is self-contained.

The trembler coil had a constantly-trembling make-and-break switch on the primary to induce high-tension electricity in the secondary. Preferable was the contact breaker, which produced a hefty pulse of current at the right moment, and (for multi-cylinder engines) a distributor to feed the high tension pulse to the appropriate cylinder.

The timing of the spark is important; the faster the engine runs, the earlier the spark should pass because combustion takes a finite time. Ignition was usually furnished with an advance/retard lever, relying on the skill of the driver to set it at the right position. Later systems provided automatic advance and retard based on a centrifugal system or the variations of pressure in the inlet manifold.

Some high-class cars, such as the Rolls Royce, sought the best of both ignition worlds by providing both coil and magneto ignition (with two sparking plugs per cylinder). With the engine stopped, moving the advance/retard lever almost always (according to the position in which the engine comes to rest) acts on the make-and-break switch, causing the coil to spark. It is a tribute to the engineering of those cars that they could be started, even after standing for some time, by switching on and flicking the lever to provide a starting spark.

Cooling and lubrication

Cooling and lubrication are also matters of great importance, and tend to go together because of the cooling advantages of a good lubricating system. At first, the need for cooling was hardly recognized, because cooling is counterproductive in the world of steam.

Later, cooling was by air or water. Air cooling was promoted by fins on the cylinder to increase the

Side view of six-cylinder engine and radiator. The water pump is mounted behind the fan; the fan draws air through the radiator to provide cooling especially when the vehicle is stationary.

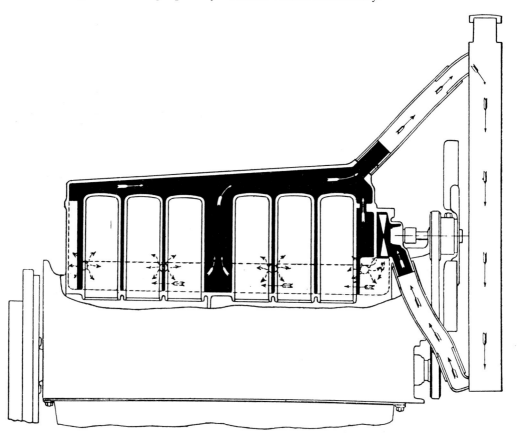

surface area (perhaps the most prominent feature of most motor cycle engines), sometimes by fans, sometimes by both. Early water cooling sometimes relied on a static water jacket round the cylinder being topped up with cold water as it evaporated. Later came the radiator – part of a closed water-cooling system – a 'heat exchanger' through which air passed to keep the water at working temperature. Water sometimes circulated automatically on the thermosyphon principle (hot water rises; cold water sinks), and sometimes it was pump assisted.

One unrelated advantage of the necessarily prominent cooling radiator was the opportunity it afforded for manufacturers to make it a distinctive shape, providing an advertisement for the marque which could be seen a mile off, in contrast to today's comparative anonymity.

Lubrication was at first drawn from steam-engine practice, for understandable reasons. As engine speeds increased, better lubrication was needed and drip-feeds and sight glasses soon became redundant as the crankshaft became enclosed in a sealed crankcase, which held an adequate supply of oil circulated either by the movement of the parts or by an oil pump.

The exhaust system

One of the most obvious sources of noise from the internal combustion engine is that caused by ignition, and expulsion of the burnt gases. It was soon found that some sort of silencer needed to be fitted to the exhaust, and that it should not impede the flow of

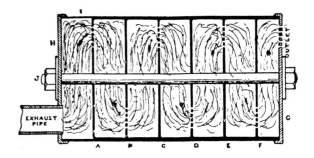

Silencers are chambers placed in the exhaust system to deaden the explosive sound of an unsilenced engine by passing the gases round a series of baffles.

gases, because back-pressure causes inefficiency. The principle of the silencer is to provide, for the cooling exhaust gases, an expansion chamber in which are baffle plates to absorb the impulses issuing from the cylinders so that the calmed gases emerge in as steady a stream as possible.

The chassis

Introduction

The coach had been a joint product of the smith, the wheelwright and the coachbuilder. In the same way, early production motor manufacturers concentrated on building a chassis to which others could add a body to the purchaser's specification. The chassis therefore had to be a rigid frame, furnished with axles and wheels, suspension and brakes, into which the engine and driveline could be dropped. Add a seat, and the chassis provided an exciting vehicle in its own right.

Front axle and steering

The standard chassis has a front (steering) axle and a rear (driving) axle, because of the problems of driving wheels which steered. Few cars had turning front axles after the manner of a cart; the danger of overturning was too great.

Her Royal Highness the Crown Princess of Roumania at the wheel of C S Rolls's 10HP Panhard (*c* 1902).

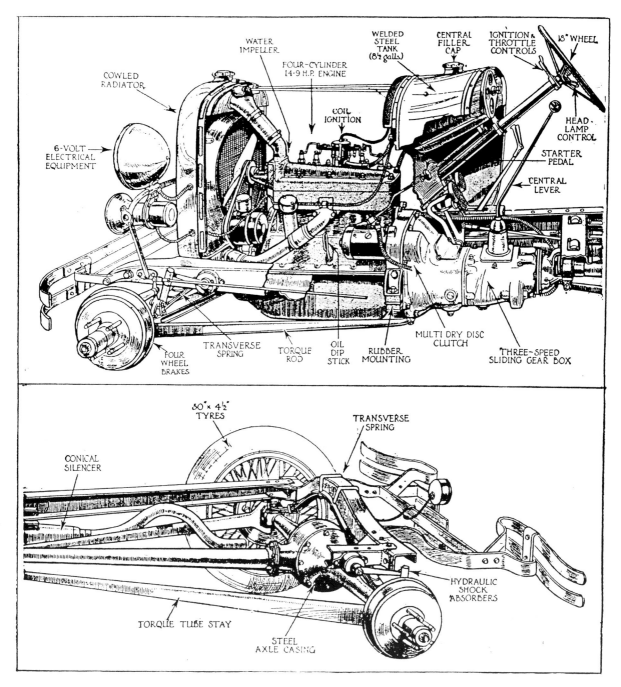

Diagrammatic views of the 1928 Ford chassis.

The front wheels were usually mounted on stub axles pivoted on king pins with steering arms connected by a track rod. Ackermann steering geometry ensures that the wheels follow the correct course when turning corners, the inside wheel being turned more sharply than the outside one.

Originally by tiller, steering was soon controlled by a wheel, giving the driver greater control. Several systems based on the rack and pinion, or the worm and sector, actuated the front wheels via drop arm and drag link.

Rear axle

When a vehicle goes round a corner, the inner and outer wheels have to turn at different relative speeds. A primitive method is to have one wheel driven and the other free, but this is not very efficient – especially if the driven wheel starts to slip. A more pleasing method is the differential gear, which transmits power to both wheels, even if they are turning at different speeds.

Driveline

Early connections between engine and driving wheels

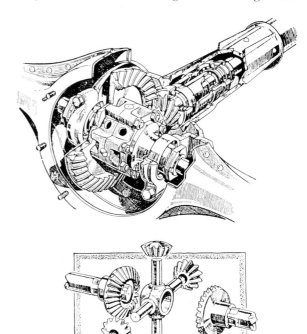

Above

The longitudinal drive of the propellor shaft is changed to a transverse drive for the rear axle by means of a bevel pinion on the end of the shaft, meshing with the crown wheel, clearly seen in the upper diagram. The crown wheel carries a cage in which small bevel pinions (shown in the lower diagram) transmit drive to bevel gears on the two half shafts of the rear axle. The two wheels will be driven positively, even if they are turning at different speeds – when cornering, for example.

Left

Rudolph Ackermann patented his steering gear in 1818 (no 4,212). The geometry is such that lines drawn though the axes of the front wheels always meet at a point (*O*) which lies on the axis of the rear wheels.

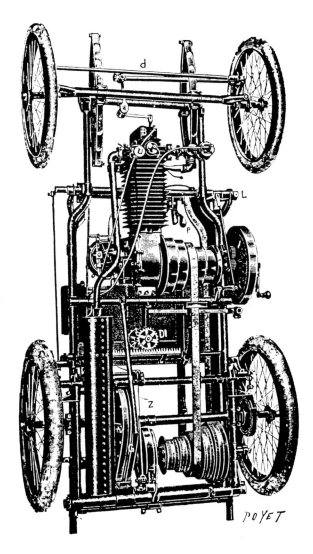

were many and various, and often drew on steam and machine-shop practice. The difference between steam and internal combustion engines is that the former can be started gradually, moving the vehicle from rest as more steam is admitted, reducing the quantity as the vehicle reaches speed, while the latter tends to generate its optimum power over a more narrow range of running speeds.

The motor car therefore has need of a clutch (to disconnect the engine from the road wheels) and a gearbox (to enable the engine to run at an acceptable speed while the vehicle starts from rest).

Early cars used a flat leather belt driving a 'loose' pulley on a shaft gradually shifted sideways on to a parallel 'fast' pulley to transmit power to the road wheels. Many different arrangements were tried, some with multiple belts to give a range of speeds.

Change-speed gears were soon introduced, so that a proper clutch became necessary; this was often of the 'cone' type, where a friction cone on the driven shaft is let into a mating cone on the engine shaft. This was felt to be more positive than the disc or plate type, although these were perfectly satisfactory as soon as suitable friction materials had been developed.

The shaft from the clutch drives the main shaft of the gearbox, wherein a positively-driven dog clutch sliding on a shaft provides a straight-through drive – top gear. Parallel to the main shaft is the layshaft, on which are gears of suitable diameters, engaged (one at a time) by sliding gears on the main shaft to give intermediate

In this underside view of the Darracq-Bollée motor carriage chassis, one is immediately struck by the stepped pulleys of the gear-change mechanism. The driving shaft is an extension of the crankshaft of the engine; the layshaft at the rear is arranged so that, as the belt is moved sideways from one pair of pulleys to another, its length may remain constant.

To transmit its drive, the cone clutch has a cone covered with friction material on the front end of the propellor shaft, kept in contact with a mating cone on the flywheel by means of a spring.

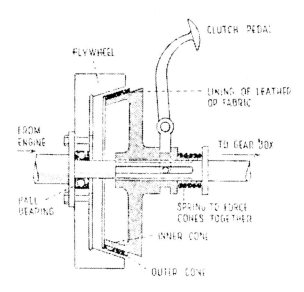

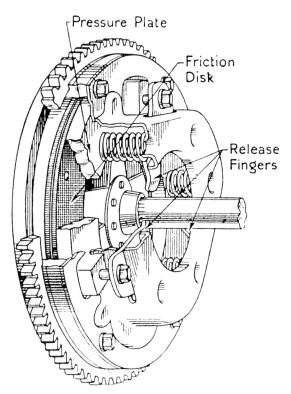

Pressure Plate

Friction Disk

Release Fingers

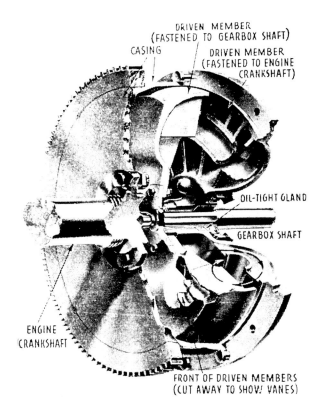

CASING

DRIVEN MEMBER (FASTENED TO GEARBOX SHAFT)

DRIVEN MEMBER (FASTENED TO ENGINE CRANKSHAFT)

OIL-TIGHT GLAND

GEARBOX SHAFT

ENGINE CRANKSHAFT

FRONT OF DRIVEN MEMBERS (CUT AWAY TO SHOW VANES)

In the plate, or disk clutch, the flywheel carries a spring-loaded plate; drive is transmitted to the disk when it is sandwiched between flywheel and plate.

In the fluid flywheel, the drive is transmitted by oil which – believe it or not – locks driving and driven members together as the speed increases.

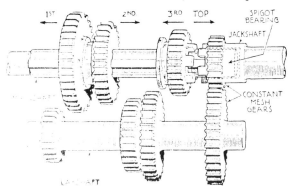

1ST   2ND   3RD   TOP   SPIGOT BEARING

JACKSHAFT

CONSTANT MESH GEARS

LAYSHAFT

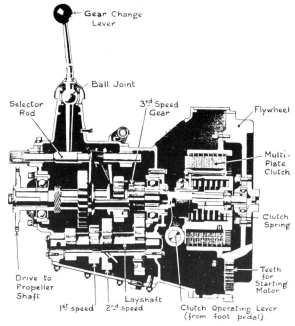

Gear Change Lever

Ball Joint

Selector Rod

3rd Speed Gear

Flywheel

Multi-Plate Clutch

Clutch Spring

Teeth for Starting Motor

Drive to Propeller Shaft

1st speed   2nd speed

Layshaft

Clutch Operating Lever (from foot pedal)

A simple arrangement of gears to provide four ratios. The small gear (top right) drives the (lower) layshaft all the time. The layshaft transmits its drive back to the mainshaft when the gears – which are locked to it radially, but not longitudinally, by means of 'splines' – are slid into engagement. Top gear is selected by sliding the gear second from the right to the right, when the teeth of the 'dog clutch' engage and the drive is transmitted straight through.

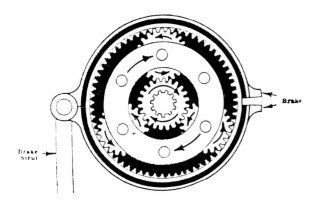

Principle of the epicyclic gearbox. The central 'sun' gear turns, and in so doing turns the three 'planets'. If the outer gear (annulus) is free to rotate, it will rotate while the planet carrier remains stationary. If the outer annulus is braked, the planet carrier must turn.

Perspective view of the 3-speed and reverse Ford epicyclic gearbox.

ratios, using the gear lever. An idle wheel provides reverse.

Because the gears have to slide into mesh while rotating, they must have straight teeth, which tend to be noisy. Changing 'crash' gears is an art; the teeth to be meshed must be moving at about the same speed. Changing 'up' is relatively easy; changing down demands 'double declutching' (depress clutch, shift into neutral, release clutch, dab accelerator to speed up layshaft, depress clutch, shift gear, release clutch) to avoid noise and embarrassment. When the synchromesh gearbox (which arguably takes much of the fun out of motoring) was introduced, the crash box was dubbed 'scrunchomesh'.

The other approach to silent gear changing was the epicyclic gear box whose gears are always in mesh, and whose ratios are selected by brake bands. It was a step from there to the preselector box, where the driver decides in advance which gear will be needed next, and changes by depressing the 'clutch' pedal.

The gearbox output is taken to the differential by means of bevel gears (see above), or a worm and

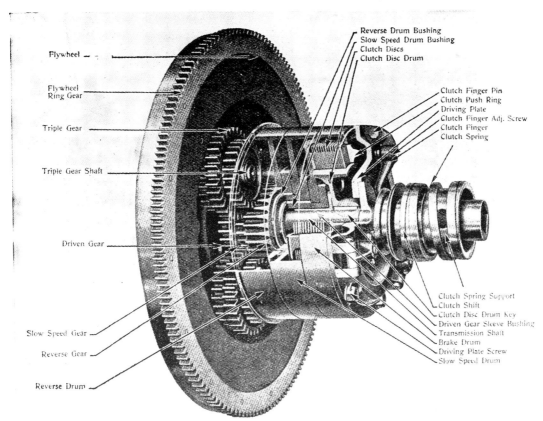

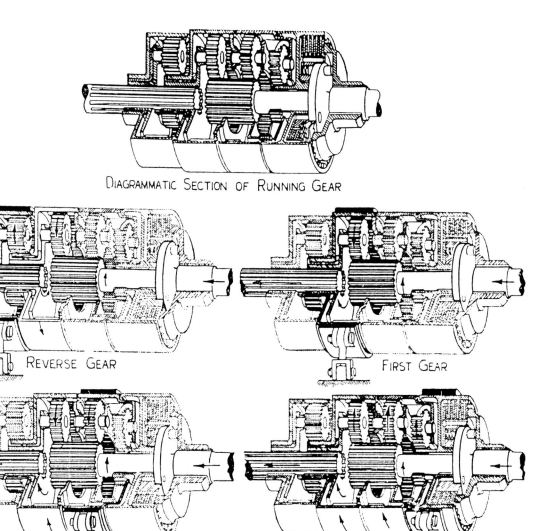

DIAGRAMMATIC SECTION OF RUNNING GEAR

REVERSE GEAR

FIRST GEAR

SECOND GEAR

THIRD GEAR

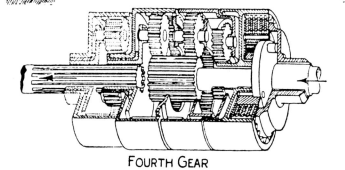

FOURTH GEAR

The epicyclic gearbox consists of a series of sun, planet and annulus sets as in the previous figure. They are so connected that the box provides a series of ratios by applying the brake bands in turn.

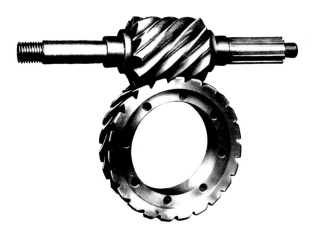

Worm-and-wheel final drive. In a rear-wheel-drive car, an underslung worm enables the floor to be lower than with a bevel drive.

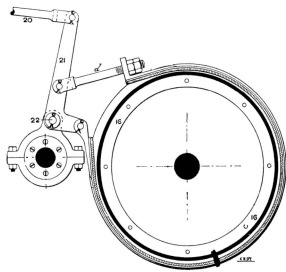

An external contracting band brake: the friction material is pulled against the drum.

wheel. The final drive may be via chains, or direct; sometimes external or internal spur gears are used to drive the wheels.

Suspension

Drawing on coachbuilding practice, most vehicles had leaf springs arranged in quarter, half, three-quarter or full elliptic configuration. Coil springs were largely avoided until they could be satisfactorily manufactured.

Sprung axles bounce about uncontrollably unless provided with shock absorbers; these afforded an outlet for much ingenuity, but the basic need was the same – to allow the springs to cushion the vehicle while damping out the bounce.

Brakes

Once there is no horse to respond to the cry of 'Whoa!' brakes become one of the most important features of the vehicle.

First came the sprag, a spike which could be let down when ascending hills to prevent the vehicle from running backwards. However, running back and jumping the sprag – the basis of many an early motoring story – could be uncomfortable, not to say dangerous.

Wood block brakes, acting on the iron-shod wheels, were soon introduced from horse-drawn vehicle practice. However, as vehicles began to travel faster, and

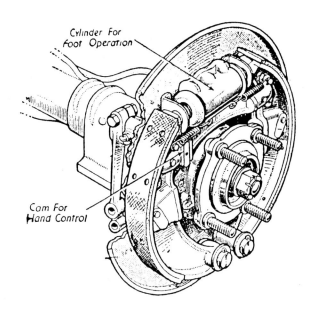

A pair of expanding brake shoes, applied by moving plungers in the hydraulic cylinders above. A brake drum fits over this mechanism, and the wheel is then bolted into position.

John Boyd Dunlop in the evening of his life, slitting open his first pneumatic bicycle tyre – which he had made himself – for display at the Royal Scottish Museum.

Little Johnnie Dunlop on his bicycle – the first to be fitted with pneumatic tyres.

One solution for the punctured tyre – the detachable Stepney rim

An 'artillery' wheel of pressed steel.

A 'wire' wheel, showing method of attachment. It was important that the fastening should not come loose while the vehicle was in motion.

Motor vehicle by Gotthold Langer of St Louis, Missouri (1897), with self-inflating tyres, and huge rear wheels so mounted that the passenger enters through them.

iron tyres were supplanted by solid rubber and then pneumatic (air-filled) ones, the need for better brakes resulted in external contracting bands acting on the propeller shaft or drums on the rear wheels.

Later, internal expanding shoes were introduced, first on rear wheels only; then on all four wheels. Brakes were controlled first by simple rods; then compensating devices were added to make braking more even; finally hydraulic systems (about 1940) provided the basis of present-day practice.

Disc brakes were not introduced until about 1960, though the pioneer motor engineer F W Lanchester had patented the scheme (no 26,407) in 1902.

Wheels and tyres

Early wheels were produced by wheelwrights using traditional techniques, and fitted with iron tyres. Wooden spoked wheels were still the norm as solid, and then pneumatic, tyres came into use. However, the pneumatic tyre introduced the possibility of the puncture, a nightmare eased by detachable rims and spare wheels. However, as both road surfaces and tyre manufacture improved, the nightmare became less frequent.

Spoked wheels were introduced as lighter and more resilient – particularly for more sporty cars – and pressed steel artillery wheels and disc wheels were also widely used.

Early private cars

Steam cars again

One of the problems with the steam-driven vehicle was the enormous bulk and weight which had to be accommodated and dragged along before any thought could be given to passengers. The coal-fired pot boiler borrowed from railway locomotive practice was the norm until the Frenchman Léon Serpollet invented the flash boiler in 1889. Here, a stack of flat coils of nickel-steel tubes was arranged over a petrol or paraffin burner, providing instantaneous superheated steam from the outlet when water was pumped into the inlet. Serpollet's developed vehicle had a four-cylinder horizontally-opposed engine with poppet valves. In 1889 Gardner, an American financier, joined Serpollet and

The 'latest type' of Serpollet steam carriage, 1897.

the Gardner–Serpollet became the steam 'Rolls Royce' of its day – on both sides of the Channel – until Serpollet died in 1907.

In 1897, the Stanley Brothers of Newton, MA, built a light (7 cwt (355 kg)) two-seater steam car with a vertical twin-cylinder engine driving the rear axle via a chain and differential. The vertical boiler heated by a petroleum burner was fitted beneath the seat, and the feed-water tank was in the boot (trunk). The car had full-elliptic springs, wire wheels and pneumatic tyres. In 1899 the Stanley brothers sold their design to the Locomobile Company of America. However, they appear to have had steam in their blood, for they rose again and produced a new and improved Stanley Steamer in 1901. The Stanley brothers were undoubtedly the founders of the American steam-car industry.

There might have been some future for the light steam car, had it – and the flash boiler – come earlier. As it was, road vehicles driven by internal combustion engines showed promise well before the end of the century, and the lightness and convenience of their power units spelt doom for the steam-powered vehicle.

Early vehicles with internal combustion engines

Etienne Lenoir

To exploit his successful gas engine, Etienne Lenoir in France set up 'Société des Moteurs Lenoir', though the engines themselves were built by Gautier et Cie, to whom Lenoir had been a consulting engineer. Although Lenoir's original engines were very large and inefficient, he gradually improved the design until, by 1862, he had a gas engine small enough to power a vehicle. Having built the car, it took Lenoir time to pluck up enough courage to take it out on the roads of Paris; in September 1863 he made a 12 mile (19.3 km)

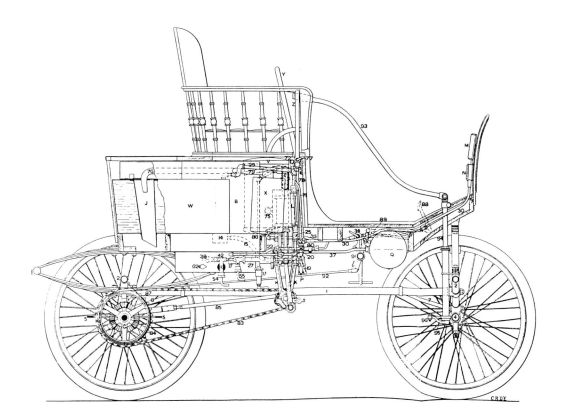

The Stanley light steam car: elevation.

journey in three hours. The 1.5 hp motor drove one of the two rear wheels via a chain. At the front, a vertical shaft carried a single wheel on full-elliptic springs; the upper end of the shaft was fitted with a small steering wheel.

Siegfried Marcus

An Austrian, Siegfried Marcus, is said to have built a vehicle driven by an internal combustion engine in about 1864. Since the flywheels of the engine served also as the wheels of the cart on which it was mounted, it is hardly surprising that it was discarded as unsatisfactory after its maiden journey. However, later research showed that the handcart had not been made until 1870, and the 'car' was built in 1888 by a company in Adamsthal.

Marcus is said to have built three other vehicles, the earliest in 1874; the later two are lost. One of his vehicles had a four-stroke engine with a horizontal cylinder whose piston drove the crankshaft via an oscillating beam rather than direct. It had a cone clutch

and belt drive to the rear axle, enabling it to travel at a brisk walking pace. It had very slow-acting steering via a worm acting on the swivelling front carriage. The fuel mixture was thrown up by a revolving brush which sprayed it into the induction pipe, and a low-tension magneto provided the spark ignition. The car is said to have travelled at 5 mph (8 kph). Marcus appears to have had little entrepreneurial stamina, for he had no interest in perfecting his designs once he had shown that they had worked. It is not known what became of his two later vehicles.

Edouard Delamare-Deboutteville

In 1883, Edouard Delamare-Deboutteville, the son of a cotton-mill proprietor, took an 8 hp stationary gas engine, adapted it for use with the more convenient petroleum spirit, and mounted it on a four-wheeled vehicle. His idea was to build a vehicle to move goods from his father's factories to the railhead at Rouen. Unfortunately, the vibration and power of the motor were too much for the vehicle and it shook itself to pieces. Undeterred, Deboutteville built a motorized tricycle, but the frame collapsed under the weight of the engine. Fortunately, he had some success developing stationary engines and won a number of awards, culminating with the Légion d'honneur in 1896.

Hammel and Johanson

The Danes Hammel and Johanson built a car in 1886 powered by a twin-cylinder horizontal engine with surface carburettor and hot-tube ignition. The rear axle is driven by a single chain from gears and friction clutches. There is no differential, but a leather-lined cone clutch on each rear wheel allows a certain amount of slip. A drive from the camshaft provides a half-speed reverse. The car has a steering wheel, unusual in 1886. However, the effect of this advance is negated by its working the 'wrong' way – turn the wheel one way, and the car goes the other.

Carl Benz

In 1885, Carl Benz of Mannheim, Germany, built a light three-wheeled vehicle running on liquid fuel. It had a horizontal water-cooled single-cylinder 0.75 hp petrol engine with a surface carburettor and coil-and-battery ignition. Vapour from the carburettor passed through a mixer – a finely perforated pipe surrounded by a larger perforated tube to admit air from the atmosphere. A sliding shutter over the outer pipe of the mixer acted as a choke.

The high-tension ignition used a trembler coil – that is, one with a constantly-buzzing make-and-break on its primary – to induce a continuous series of pulses of energy in the secondary coil. A rotary switch (distributor) directed energy to the sparking plug. The ignition could be advanced and retarded by turning the rotary switch about its axis – a principle used ever since.

The engine had a vertical crankshaft and hence a horizontal flywheel – so mounted to overcome its gyroscopic effect, real or imaginary. The power was transmitted via bevel gears from the top of the crankshaft to a short horizontal shaft carrying fast-and-loose pulleys from which a belt drove a countershaft placed low down in the centre of the car. The countershaft incorporated a differential gear, and drove the rear wheels via chains. The single front wheel was steered by means of a tiller via a rack and pinion. This first Benz motor travelled at about 9 mph (15 kph), and was demonstrated in public on 3 July 1886.

Over the next two years, Benz built a 1.5 hp and then a 2 hp car which was awarded a Gold Medal at the Munich Industrial Exhibition in September 1888. This inspired Benz to build further similar cars – without much marketing success until the Parisian dealer Emil Roget came to the rescue. Hitting on the important difference between manufacturing and marketing skills – perhaps unwittingly – Benz and Roget formed the world's first alliance between a motor manufacturer and a dealer.

Two views of the 1885 Benz motor tricycle. Note the exposed crankshaft with flywheel below and gears above. Fast-and-loose pulleys are seen to the right below the bench seat.

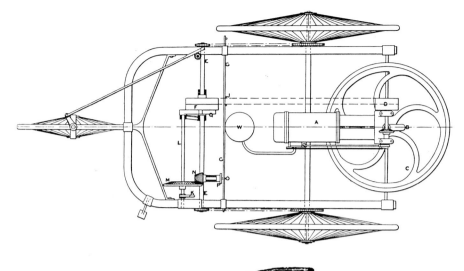

Simplified plan view of the first Benz motor tricycle. The fast-and-loose pulleys drive a countershaft (E–E), from which chains drive the rear wheels.

Benz's second motor tricycle carriage of 1886. Its maximum speed was 10mph.

Benz's production vehicles had two 4ft 6in (1.37 m) diameter rear wheels with iron tyres and a single 2ft 6in (0.76 m) front wheel with a solid rubber tyre mounted on a fork and steered by rack and pinion. Braking is by wooden blocks operating directly on the tyres of the rear wheels. The single-cylinder water-cooled engine, with mechanically-operated overhead inlet and side exhaust valves and a vaporizer carburettor, is mounted over the rear axle. Ignition is by coil and sparking plug. There is no radiator; the driver has to top up the water reservoir at the rate of about one gallon (4.5 l) every hour. The flywheel is on the lower end of the vertical crankshaft; at the upper end is a

bevel gear driving a cross shaft with a belt drive to fast-and-loose pulleys on the main countershaft, which drives the two rear wheels via chains and differential.

In 1893, after further development, Benz built a four-wheeled vehicle with a 3.5 hp motor with a vertical flywheel and two-speed belt drive, and by the end of that year had sold 69 vehicles in all. He continued to produce vehicles based on his 1893 design until 1901.

Gottlieb Daimler

The cars described so far had low-speed engines, as a result of which the drive to the wheels had to be fairly

Benz's third motor carriage of 1888 ran at speeds up to 15mph (24 kph). It has a steering wheel, and is not underpowered.

direct. In the mid-1880s, Gottlieb Daimler of Cannstatt, Germany, experienced in building gas engines, saw the advantage of a high-speed internal combustion engine, and developed a simple, light, vertical single-cylinder engine running at hundreds of revolutions per minute. In British patent no 9,112 of 1884 he describes a high-speed gas engine with the bore small in proportion to the stroke of its piston, running at sufficiently high speed to ignite the charge by the heat of compression, aided by an ignition tube to maintain regularity. In patent no 4,315 of 1885, he describes a single-cylinder engine with enclosed crank and flywheel, and automatic inlet and rod-activated exhaust valves. A governor prevented the exhaust valve from opening when a set speed was reached.

In 1885, Daimler designed a surface carburettor enabling the engine to run on light petroleum spirit. Daimler's carburettor has a vessel kept about two-thirds full of petrol, with an annular float bearing a tall vertical tube with perforations near its base. Above the main vessel, a smaller chamber provides a reservoir for the air–petrol mixture. The engine draws its mixture from this chamber, and the reduced pressure bubbles more charged air from the lower vessel.

Daimler mounted his engine on a specially-built bicycle in 1886. In the same year, he mounted one of his engines on a four-wheeled horse-drawn carriage, removing the shafts and adding a steering mechanism. However, he was one of the first to realise that there was little future in the 'horseless carriage', if that was how such vehicles were viewed. Engines should not be mounted in existing vehicles; the vehicle should be designed as a whole. Daimler's Stahlradwagen (steel-wheeled carriage) of 1889 was the first vehicle to be designed on this principle. It was also the first to use Daimler's two-cylinder (comparatively) high-speed V-engine, and travelled at 11 mph (17.5 kph).

The two cylinders of Daimler's engine (1889 patent), were inclined to one another at about 15° with their connecting rods on a common crank. Daimler's earlier air-cooled engines had fans; this model had water cooling and a radiator. He sold these engines to other car makers; they were also used in boats and as stationary engines.

In 1889, Daimler built a successful quadricycle (four-wheel cycle), followed by many other four-

Gottlieb Daimler's motorcycle of 1886.

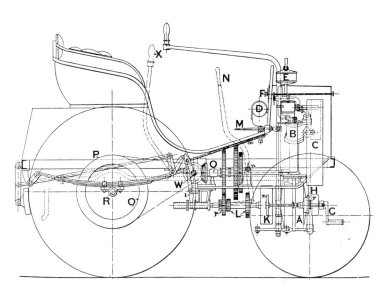

Messrs Panhard & Levassor acquired the rights to Daimler's engine in 1889, and immediately began to build motor cars. This 1894 model is very exposed – particularly the gearbox.

wheeled motor carriages of various designs. From 1893, he used Wilhelm Maybach's patented float-feed carburettor. In Maybach's design, petrol entered a float chamber, causing the float to rise. When the float had risen to a predetermined height, it cut off the supply, thus maintaining a constant head. The petrol from the float chamber was fed to a jet, whence the suction from the engine caused it to emerge and evaporate. This principle has formed the basis of carburettors up until today.

All Daimler's engines had hot-tube ignition and belt transmission and from 1894 were built at his Cannstatt Works. Cannstatt-Daimlers have nothing to do with the English Daimlers, which were built at Coventry from 1896.

Benz (1844–1929) and Daimler (1834–1900) never met, but their two businesses came together as Daimler Benz in 1926. The firm continued to manufacture the Mércèdes, first named after Daimler's daughter in 1901.

Edward Butler

The earliest English vehicle with an internal combustion engine was patented by Edward Butler in 1887. Butler called his motor tricycle (which he built and tested in 1888) a 'petrol cycle' – the first recorded use of

the word 'petrol' (American 'gas').

It had a horizontal twin-cylinder engine with a direct drive to the single rear wheel, whose axle was the crankshaft. However, the engine did not seem fast enough, so Butler added gears to give the vehicle a more satisfactory turn of speed. Because the drive was direct, the machine was raised off the ground before starting the engine; dropping it must have needed some courage. The two front steering wheels of Butler's tricycle were controlled by side levers. But, just as the tolls had driven the steam coaches off the road, so did the legal speed restrictions discourage Butler from further development.

Peugeot and Panhard

The first petrol-driven car built in France was produced in 1889–90 by Armand Peugeot. Peugeot used a rear-mounted V-twin Daimler engine built under licence by Panhard et Levassor of Paris. In 1896, Peugeot built his own horizontal twin-cylinder engine, and continued to produce rear-engined vehicles until 1902.

In 1889, the French firm of Panhard et Levassor built an experimental car, having secured the French manufacturing rights for the Daimler motor. In 1891, they built a car with a vertically-mounted twin-cylinder Daimler engine at the front, a friction clutch and a

The Peugeot 'Benzine' car of 1895.

John Henry Knight was interested in road vehicles rather than in the means of propelling them. Before he turned to building petrol vehicles, he produced this 32 cwt (1,625 kg) steam carriage in 1868, and made several trips in the autumn of that year. The vehicle was so often altered and modified that little remained of the original.

three-speed gear box. Bevel gears at the rear of the gear box drove a countershaft, whence a chain in the centre of the vehicle carried the power to the rear axle – which was fitted with a differential gear.

The 1891 Panhard layout set the standard for most motor cars for some time to come. It had Ackermann steering, though it was controlled by a tiller. Later, Panhard was one of the first manufacturers to adopt the steering wheel.

John Knight

John Henry Knight of Farnham in Surrey seems to have been one of the few pioneers wedded to the quest for a satisfactory road vehicle rather than to a method of propulsion. He built his first steam road vehicle in 1868 and in 1895 built a three-wheeled petrol-driven car with a horizontal single-cylinder engine, fitted first with hot tube ignition, then with electrical ignition. It had two-speed belt transmission, the belt in use being tightened by a jockey pulley, while the other remained slack. (The drive was in neutral when both belts were slack.) This experimental vehicle was used as a test bed

for various modifications; it was later converted to a four-wheeler, for example.

Frederick Lanchester

F W Lanchester (1868–1946) was a well-known engineer who turned his mind to motoring. As an engineer, he saw – like Daimler before him – the need for a complete review of motor-car practice – if only to get away from the idea of the 'horseless carriage'. Lanchester completed his first design in 1896; the frame of his machine was of brazed steel tubes, and an inclined 5 hp single-cylinder air-cooled engine with a wick carburettor was mounted in the middle. To reduce vibration, Lanchester's engine had two contra-rotating crankshafts geared together, each with its own flywheel. One of the crankshafts drove the live rear axle via a chain, epicyclic gearbox and differential gear. The gearbox provided low and high speeds (direct drive) and reverse. The wire wheels were fitted with Dunlop pneumatic tyres; the car had tiller steering.

Lanchester then designed an 8 hp horizontally-opposed twin-cylinder air-cooled engine – again balanced with twin flywheels, one of which incorporated the low-tension magneto ignition components. The inlet valves were mechanically operated. This engine was mounted at the rear of the chassis of his new vehicle, driving the rear axle via an improved epicyclic gear

and worm drive. This Lanchester ran at 20 mph (32 kph) in 1897.

Lanchester's experimental car of 1898 attained a speed of 28 mph (45 kph) on the level and made many long journeys without breakdown. In 1899 the car was awarded a gold medal for its performance. Later in the same year, Lanchester set up the Lanchester Engine Company to produce a 10 hp car similar to his gold-medal winner; it sold well until 1905, when he designed and built a more conventional engine.

Herbert Austin

Herbert Austin (1866–1941) designed his first motor vehicle while working for the Wolseley Sheep Shearing Machine Company in 1895. It had three wheels, a tubular-steel frame, and a horizontally-opposed 2 hp twin-cylinder engine mounted on the nearside of the frame. The crankshaft was extended to pass through the hub of the single rear wheel to the flywheel. A flat-belt pulley drove a three-speed gear box (beneath the driver's seat) whose output returned to drive the rear wheel via a chain.

Austin's second three-wheeler of 1896 had a single front wheel steered by a long tiller, and a small dog-cart body in which driver and passenger sat back to back. The first engine – a twin-cylinder water-cooled unit with epicyclic transmission – was unsatisfactory;

An early experimental motor car, built by Herbert Austin in about 1896.

A Wolseley 8 hp car – available with standard body for £175!

Austin replaced it with a single-cylinder horizontal engine with air and water cooling. Belt transmission provided two forward speeds via two belts and fast-and-loose pulleys and a chain drove an axle via differential and reduction gearing. Spur gears at the end of each half shaft drove the rear wheels. The vehicle had independent rear suspension: each wheel could move independently in an arc against a coil spring.

The first four-wheeled Wolseley car appeared late in 1899. It had a horizontal single-cylinder water-cooled 3.5 hp engine with detachable cylinder head, a front-mounted finned-tube radiator, trembler-coil ignition, and a single-jet carburettor. A flat belt drove the three-speed gear box, which drove the rear wheels via a chain. The gear lever acted also as a clutch; moving it

sideways selected the gear and pushing it forward tightened the belt drive. The front wheels were steered by a tiller and Ackermann system. The car had a two-seater body and pneumatic tyres, and performed well in the 1900 Automobile Club of Great Britain and Ireland 1,000 miles trial, winning several awards.

Henry Ford

In America, Henry Ford (1863–1947) became interested in road vehicles in about 1890 and produced his first car in 1896. Sometimes referred to as a quadricycle, Ford's original car had a twin-cylinder four-stroke water-cooled engine, belt transmission, tiller steering, wire wheels with solid rubber tyres and is said to have attained a speed of 30 mph (48 kph). Ford's second car, built a few years later, was even more successful. He founded the Ford Motor Company in 1903 and produced one of the most successful models ever; 15,000,000 Ford Model T cars were built between 1908 and 1927. Ford's contribution to the industry was to bring the motorcar within almost everyone's reach by using mass-production techniques pioneered by the arms manufacturer Samuel Colt (1814–1862) to provide interchangeability. This avoided expensive and time-consuming selection and fitting of parts, which not only kept the price of the vehicle down, but made repairs and servicing easier and cheaper too.

With only one model (the Model T) to build – in 'any colour you like, as long as it's black' – the Ford plant at Detroit, Michigan produced over a million cars in 1914, at an unheard-of price of $600.

Renault frères

The Renault brothers lived at Billancourt, France. Towards the end of 1898, Louis Renault, dissatisfied with the performance of his 1.75 hp de Dion tricycle, fitted its engine into a four-wheeled car of his own design. This light vehicle had a three-speed gear box driving a differential gear in the rear axle via a shaft. He tried to patent his design in 1899, but his application was quashed both in France and in England. Nevertheless, the whole of the vast Renault business was built on this original car with its small air-cooled engine.

Henry Ford with his wife and grandson (Henry Ford II). Henry Ford sits at the tiller of the first car he made, in 1896 when he was 33.

The Tin Lizzie, or Model T Ford – a four-seater type of 1910.

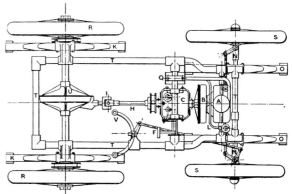

Plan view of chassis of Renault's light motor carriage of 1898, with single-cylinder 2.25HP de Dion & Bouton motor.

On the road

The Hon Charles Stewart Rolls (1877–1910) was the third son of the 1st Baron Llangattock. Born in London, he was educated at Eton (where he specialized in practical electricity) and Trinity College, Cambridge (where he read mechanical engineering and applied science). He was also a prominent cyclist, and, by way of a change, built a ten-seat tandem. After Cambridge, he went to work in the London and North Western railway works at Crewe and also took a third engineer's (marine) certificate.

In December 1895, Rolls became the fourth person in England to own a car – a Peugeot. In those days, the speed of road vehicles was limited to four miles an hour and they had to be preceded by a man carrying a red flag. The 'Red Flag Act' was repealed in August 1896, and the top speed of 12 mph (19 kph) was raised to 20 mph (32 kph) in 1903.

Frederick Henry Royce (1863–1933) started work at the age of ten when his father died. He had a series of jobs, teaching himself to become an engineer by diligent study at night school, and was appointed chief electrical engineer for the London Light and Power

Capt the Hon C S Rolls, MVC (*c* 1903).

Sir Henry Royce (1863–1933).

Rolls driving HRH the Prince of Wales (later King George V) on his 12HP Panhard.

Company in 1882.

Royce started his own electrical engineering works in 1884. It was not until 1903 that he bought a motor car – a 10 hp Decauville – which was so noisy and unreliable that he resolved to build a car of his own. He designed every detail with meticulous care and drove the first 10 hp two-cylinder Royce from his factory on 31 March 1904.

C S Rolls, who by this time had set up a London dealership for the finest continental cars, was so fired with enthusiasm when he encountered the 1904 Royce that he arranged to buy all Royce's cars and sell them under the name of 'Rolls–Royce'. The two firms combined in 1906. In the same year, Rolls set up a record for the Monte Carlo to London run in a 20 hp Rolls–Royce, covering the 771 miles in 28 hours 14 minutes – a speed of 36.3 mph (58.4 kph). Deciding to concentrate on a single, reliable model, Rolls–Royce brought out the Silver Ghost, a six-cylinder model which was produced with little change until 1925.

Rolls also made 150 balloon ascents and took up aeronautics as soon as the first frail flying machines reached England. He won many trophies and in 1910 crossed and recrossed the English Channel in 95 minutes – a record for the time. His final, unfortunate,

'first' was to be the first Englishman to be killed in an aeroplane crash when the tailplane of his machine collapsed at a display at Bournemouth on 12 July 1910.

Rolls published his 'touring reminiscences' in 1904. He began by quoting the words of a friend of his:

Here's a nice position for me to be in! There's the motor that's killed Rolls; it's running away, and I don't know how to stop the d–d thing, and there's Rolls like a sack of bones – dead as mutton – and I've got to tell his people that he's been run over by his own car! I don't know them; and I don't even know where they live!

Rolls continues:

My car was standing in an hotel yard at Gloucester; the engine was running, and I noticed that the clutch pedal stuck down. I stooped in front of the car to release it, but carelessly left the gear lever in the first speed. The result was that as soon as the pedal flew up the car started ahead, and my friend had the pleasure of seeing me, after an ineffective struggle to get out of the way, lying flat on the ground with the car passing over me. The conclusion that I was dead was perhaps justifiable, for I must have presented a somewhat pitiable spectacle with my clothes dragged away by the projecting parts of the engine, and the cape of the old macintosh I was wearing drawn over my head. For the moment, indeed, I was not altogether certain myself whether I had come through on the right side of eternity; but I soon made up my mind. Peeping out I discerned the car careering down the road in the

The fourth car-owner in England – Rolls on the 3.75HP Panhard, built in 1896.

direction of a dogcart, and my resurrection took place just too late to prevent a smash.

Not a very auspicious occasion in the motoring career of one whose name was to become synonymous with quality. Stepping out of the car with the gear engaged, the engine running, and the clutch stuck down was fairly unwise. One would have thought that an engineer would have known better! Rolls continues:

This incident was one of a series in what was the first really exciting tour I ever had. It was at Christmas time, 1896, and I set out to travel from London to my home in Monmouthshire. After innumerable difficulties, including not a few breakdowns, we passed through Cirencester, and had to descend Birdlip Hill. Even now [1904], with the latest improvements in brakes, Birdlip Hill is one of the most hazardous inclines in England, and at that time, with the embryonic mechanism at our command, it was a hill that might have given pause to the most reckless of chauffeurs. Some idea of its dangers may be gathered from the fact that between the highest point at Birdlip and the comparatively level going of Ermine Street it drops something like 700 ft in a mile and a half [about 1 in 11].

I am bound to say that had we known what the route was like we should never have attempted it; but we didn't know. The road, with its curves, its corners, and its bumps, was new ground to us, and at it we went hands down. I have never tried to slide down a precipice, but I should imagine the sensation would be something akin to ours as we began that descent. We started cautiously, as people should who slide down precipices, and just as we reached the steepest part I

suddenly felt my hand-brake lever yield and go up against the stop. A feather had sheared, and there was an end to the side brakes.

A feather, or key, in this case would secure the hand-brake lever or other part to a shaft. If the feather shears, there is no way of applying force to the hand-brake system. Rolls continues:

We had then to rely on the foot-brake, and I pressed this so hard that the lever bent, and the pedal rested on the floor. That relieved the car of all hindrances, and away she went. There was nothing for it but to keep her in the middle of the way, and trust to luck; and it seemed a precious poor outlook. With every foot the car gained speed, leaping terribly from side to side, while the hedges on either hand flickered past like the pictures in a biograph.

'Then came a corner. The speed at which we were going left me very little breath to spare, but I certainly breathed more freely when we were safely round that angle. The car literally jumped at it, sprang from one side to the other, and we were round.'

Fortunately, none of the other cars in the country seemed to be in the vicinity.

The advice given to motorists on cornering was as follows:

Always keep to your right side, remembering that in all probability you will find some other vehicle coming towards you from the opposite direction. It will generally be found that as the road slopes towards the gutter, the outside wheels of the carriage will be higher than the inside. The illustration shows how, when encountering a bend or corner the view round which is not interrupted by hedges or other obstacles, a driver – being certain that there are no other vehicles or persons beyond the corner – may take advantage of the banking of the road, and avoid great deviation from the straight course, by cutting across to his wrong side, and hugging close to the angle of the corner.

This is a practice which is adopted when travelling at high speed as in races, but, of course, should never be attempted on the road, *unless it can be clearly seen that the track beyond the bend is absolutely free from vehicles or passengers*; otherwise, with the growing use of automobiles, serious collisions would ensue.

As for Rolls:

I had hoped that was the end; but not a bit of it. Ahead was a long, straight stretch, steadily descending. There was no stopping, and we took this in the same style, speeding with ever-growing acceleration to the next bend. Suddenly, as we neared it, we saw a light right in front of us, and we felt that nothing short of a miracle could save us from disaster. Nearer and nearer we drew to it in the increasing darkness, and at

How to take a corner – 1904 style.

length on reaching it we found to our intense relief that the light came from a house at the side of the road.

Somehow or other we got round this corner, and again there was a gleam of hope that we might be near the end of the nightmare. But the road still went on, and not until two more stretches had been covered, and two more bad corners negotiated, did we find ourselves at last slowing up, and then it seemed to me miles before the car actually came to a standstill. When we stopped there was a strong smell of burning, and we found that all the brake leather had been burned. After adjusting the brakes to the best of our power, we decided to put up at Gloucester.

It was at Gloucester that the incident of the jammed clutch took place, culminating in the car smashing into a dog cart. Either the account is somewhat embroidered, or the car was very strong because the journey continued:

The adventures of that journey were by no means over, though those that remained were insignificant by the side of our experience on Birdlip Hill. In Gloucester we had considerable difficulty in finding any benzoline, and when at last we drove away we had trouble with the pump and got stuck on a lonely hill with the engine seized up. To add to our embarrassments, so strong a wind was blowing that our [acetylene] lamps would not keep alight, and, as our matches were soaked with rain, we had perforce to work in pitch darkness. At length, however, we did reach Monmouth, at 3am on Christmas morning. The run of 140 miles had taken us three days.

Rolls now describes his imported vehicle:

The car on which I made this journey was a Peugeot carriage of three-and-three-quarter horse-power. As the most powerful one previously made was three-and-a-quarter horse-power, mine was regarded as a very terrible and dangerous machine. It was not particularly dangerous or terrible, but, certainly, in the light of recent improvements, it left a good deal to be desired. It was a rather top-heavy affair, hung on three springs, and it used to sway fearfully when going down hill – an accomplishment, by the way, which probably saved our lives on Birdlip Hill by checking its speed. It had a V-type Daimler engine behind, made by Messrs Panhard & Levassor. The thing created something of a sensation when I brought it to England. To begin with, on attempting to drive it out of Victoria Station I was stopped and summoned for not having a red flag. It was, accordingly, necessary to accomplish the first part of the journey to Cambridge, whither I was bound, in strict accordance with the demands of the law, and my friends and I took turns to walk in front with a red lantern. After we passed Potter's Bar we gave a policeman a lift. I happened to be carrying the lantern, and met him in the road.

"Evening" said I.

"One of these 'ere 'orseless carriages?"

"Yes," said I.

"Don't see many of them there things about, but I *should* like to have a ride on one; they does seem to me so huncanny."

"Jump up!" said I, and off we went, the bobby holding his helmet on with one hand, and himself on with the other. In due course we arrived at Royston, where, as we learned afterwards, a crowd, having heard of our approach, had waited through the night to greet us. And finally we got to Cambridge, after a run of eleven and three-quarter hours from London. Something under five miles an hour! Well, things have changed since 1896.

Rolls then goes on to describe his Bollée Voiturette:

The first time I tried a Bollée was in 1897. It was a curious-looking, low-built three-wheeler, with a single driving wheel behind. Rumour said that Bollées were constructed mostly of old biscuit-tins, and they certainly looked it. The cylinder

Rolls and friend on his 4HP Bollée racer (*c* **1900).**

wore oval in a week, and would not hold any compression, and in spite of copious oiling the back of it was generally red hot. In many ways the possession of a Bollée was a chastening experience. It taught you, for one thing, the wisdom of early rising. I discovered after one or two experiments, that if you wanted to start at ten in the morning it was necessary to begin coaxing the machine at about six.

Here is a passably true record of one of the mornings I spent after I had become fairly well acquainted with my Bollée's idiosyncrasies. Punctually at 6.15 a.m. I began lighting the burner. At 6.23 I was painfully reminded of a man at college who, having dined, mistook a hat-peg for a gas-jet and wasted a box of matches trying to light it; I had turned on the wrong tap. At 6.25 the crowd, already large for a country place, began rapidly to increase, and by 6.26 they were shouting for the fire brigade. Between 6.30 and 6.40 I delivered a vigorous and impassioned address to the excited multitude. At 6.50 I had succeeded in getting the burner to work, and by 7.05 it was making a noise like the safety-valve of a Channel steamer blowing off. Meanwhile I had been looking for the starting handle, and at 7.18 I remembered that it had been dropped on the road the day before, and that someone had been sent in search of it. At 7.27 this person, having been found, declared that to the best of his recollection he had deposited the handle in one on the cupboards of the hotel coffee-room. By 7.55 all the cupboards had been searched, and the missing article had been found in an empty pail under the stairs leading from the public bar! At eight o'clock I had finished thanking the gentleman for the trouble he had taken to put it in a safe place, and had begun the task of

winding up. At 8.55 I adjourned for breakfast, and appointed a reliable-looking groom as my deputy winder, with orders to send for me if his labours produced any appreciable result. At 9.10 a messenger in frantic haste reported a slight explosion, and I conveyed a cup of coffee and a piece of buttered toast to the seat of the car. By 9.25 there had been no further explosion, but there was a bad back-fire which nearly broke the man's wrist. At five minutes to ten the engine started, flung the handle in the face of the reliable-looking groom, and knocked into his mouth one of his front teeth – luckily, as I subsequently discovered, a false one. And at 10.10 I was fairly on the road, the machine grunting, panting, and throbbing like a cross between a pig and a traction engine. It will readily be believed that runs which began so strenuously were full of interest, and that the car, once started, required a good deal of stopping. The end of my first run on a Bollée – from Coventry to Cambridge – was suitably marked by a collision with a cart.

Rolls seems to have been particularly unfortunate with carts; there is no doubt that he was making a good story of his experiences, though none of the incidents described is in any way implausible. Difficult as it may be for us to appreciate the trials of early motoring after nearly a century, Rolls's account gives us some feeling for what it must have been like, and makes us appreciate the improvements in vehicles and roads since his time.

Chronology

Further reading

George Dammann
Illustrated History of Ford 1903–1970
Crestline Publishing, 1971

Frank Donovan
Wheels for a Nation
New York, Thomas Crowell, 1965

James Flink
The Automobile Age
Cambridge: MIT Press, 1990

Edward Janicki
Cars Detroit Never Built: Fifty Years of Experimental Cars
New York, Sterling, 1990

Leon Mandel
American Cars
New York, Stewart Tabori & Chang, 1982

Ray Miller and Bruce McCalley
From Here to Obscurity: an Illustrated History of the Model T
Evergreen, 1971

Marco Ruiz
100 Years of the Automobile
London, Gallery Books, 1985

Charles W Singer *et al*
A History of Technology
Oxford, at the University Press, 1954

Index